DEPARTMENT OF ENERGY

Energy Paper Number 13

National Energy Conference June 22, 1976

Volume II
Papers submitted

LONDON: HER MAJESTY'S STATIONERY OFFICE

First published 1976

ENERGY PAPERS

This publication is the Volume II of the thirteenth in the Energy Papers series published by the Department of Energy.

The series is primarily intended to create a wider public understanding and discussion of energy matters, though some technical papers may appear in it from time to time.

The papers do not necessarily represent Government or Departmental policy.

Other papers in the series:

Number 1 *UK oil shales: past and possible future exploitation* (£1.15)

Number 2 *Methane derived alcohols: their use as blending components of petrol* (80p)

Number 3 *Advisory Council on Energy Conservation: Report to the Secretary of State for Energy* (80p)

Number 4 *Energy: the key resource* (43p)

Number 5 *Energy saving: the fuel industries and some large firms* (£1.85)

Number 6 *Advisory Council on Energy Conservation: Paper 1: Policy statement by the Trades Union Congress* (65p)

Number 7 *North Sea costs escalation study* (£3.35)

Number 8 *The Offshore Energy Technology Board: strategy for research and development* (90p)

Number 9 *Geothermal Energy: the case for research in the United Kingdom* (£1.85)

Number 10 *Advisory Council on Energy Conservation: Paper 2: Passenger Transport* (90p)

Number 11 *Energy R & D in the United Kingdom* (£2.65)

Number 12 *Advisory Council on Energy Conservation: Paper 3: Energy prospects* (£1.10)

ISBN 0 11 410291 0

Contents

Page

v Foreword by the Secretary of State for Energy

1 Agenda; Areas for decisions; Topics for discussion; Questions

4 World energy resources: present position and prospects

15 UK energy supply and demand prospects

22 *Energy prospects (Energy Research Group, Cambridge University)

COAL

23 Long term prospects for coal (Joint Policy Advisory Committee of the Coal Industry)

GAS

27 United Kingdom energy prospects and plans for gas (British Gas Corporation)

ELECTRICITY

32 The prospects for electricity (Central Electricity Generating Board)

34 Energy and electricity in Scotland (South of Scotland Electricity Board)

38 Interconnection—its place in a national energy policy (Northern Ireland Electricity Service)

40 Flexibility in long-term energy policy (Electricity Council)

OIL

45 The role of the oil industry in UK energy supply and demand 1975–2000 (Petroleum Industry Advisory Committee)

47 Note from the British National Oil Corporation

50 UK offshore petroleum exploration and production (United Kingdom Offshore Operators' Association)

NUCLEAR

56 The future of nuclear power in the UK (United Kingdom Atomic Energy Authority)

58 Energy policy and the nuclear industry (Nuclear Power Company Limited)

ENERGY CONSERVATION

61 Energy conservation (Advisory Council on Energy Conservation)

RESEARCH AND DEVELOPMENT

64 *Energy R & D in the United Kingdom: a discussion document (Advisory Council on Research and Development)

CONSUMER GROUPS
68 Energy for industry—the industrial consumer's view (Confederation of British Industry)
76 Fuel pricing policy (National Consumer Council)
81 Statement by the Consumers' Association
83 Fuel Poverty (Child Poverty Action Group, Age Concern, and Help the Aged)
88 Towards a national energy strategy (Friends of the Earth and the Conservation Society)
POLICY
92 *Transcript of the coal and electricity industries tripartite consultations held at Church House, Westminster on Friday, February 20, 1976
93 The case for a planned energy policy (Frances Morrell and Francis Cripps)
97 TUC fuel and power policy (TUC Fuel and Power Industries Committee)
102 Energy policy—a trade union strategy (National Union of Mineworkers; Electrical Power Engineers Association; National Association of Colliery Overmen, Deputies and Shotfirers)
107 Energy policy—the case for integration and expansion (Amalgamated Union of Engineering Workers)
109 Statement on energy policy (Electrical, Electronic, Telecommunications and Plumbing Unions)
111 Energy policy (Transport and General Workers Union)
114 The layman's guide to the requirement for energy (Amalgamated Union of Engineering Workers—Technical, Administrative and Supervisory Section)
117 Labour Party
121 Conservative Party
122 Liberal Party

*These papers submitted to the Conference are published separately in the Energy Papers series.

Number 11 *Energy R & D in the United Kingdom, a discussion document*

Number 12 *Advisory Council on Energy Conservation: Paper 3 : Energy prospects*

Number 14 *Tripartite energy consultations, February 20, 1976*

Foreword by the Secretary of State for Energy

This is Volume II of an Energy Paper which fulfils an undertaking I gave at the National Energy Conference in London on June 22, 1976 to publish a complete record of the event.

This volume reproduces the Agenda and all the papers prepared for and circulated in advance of, or on the day of, the conference, except where they are being published separately in this series.

Those conference papers which are being published as separate Energy Papers are:

—Energy R & D in the United Kingdom (Energy Paper 11)

—Energy Prospects (Energy Paper 12)

—Transcript of Tripartite Coal/Electricity Meeting, February 20 (Energy Paper 14)

The first volume of the National Energy Conference record brings together the transcript of the proceedings; contributions which would have been made had time permitted me to call further speakers; and the list of those invited to attend.

Tony Benn

Secretary of State for Energy

Agenda; Areas for decisions; Topics for discussion; Questions

SESSION 1 (10.30 am—1.00 pm)

Agenda

The supply and demand for energy; UK prospects to 2000; problems and constraints; priorities and possible solutions. The role of each energy industry; its problems, plans and decisions.

Areas for decisions

Investment in coal.

Rate of depletion of oil and gas resources.

Construction of new power stations: how many, when, and what type.

Development of fast reactor.

Allocation of resources for research and development (including unconventional energy sources and conservation).

Topics for discussion

1 An energy gap?

There is general agreement that the medium term prospect for the UK energy economy is very favourable. We are fortunate in having the prospect of being able to produce in the 1980s, mainly from coal, oil and natural gas but with a substantial contribution from nuclear, energy equivalent to our needs.

How long can this situation be expected to last? What are the prospects thereafter?

Broadly there are two views:

—That the UK's energy self-sufficiency cannot be expected to last much beyond 1990 if at all, and that there is a danger that, unless vigorous forestalling action is taken, energy import requirements would increase rapidly towards 2000 at a time when energy on world markets may be both scarce and increasingly expensive.

—That future discoveries of oil and gas in UK waters may prolong self-sufficiency at least until early next century, and that, because of the existence world-wide of ample resources of coal capable of being mined at relatively low cost, and because of the prospects that by 2000 alternatives to fossil fuel, both nuclear and other, will be making a major contribution, international trade in energy will be free-moving and renewed dependence on imports for a substantial part of our needs would be quite supportable.

(Intermediate views would compound optimism about UK energy reserves and production with pessimism about developments on the world energy scene; and vice versa.)

Questions:

(a) On what view of the future should we base our planning?

(b) What are the lead times for preparing for possible scarcity around the end of the century, in terms of:

research and development.
coal production.
nuclear production.
investment in energy conservation?

(c) Is the difference between an expectation of 'self-sufficiency till early 1990s' and 'self-sufficiency till after 2000' critical for current decisions?

(d) In view of the uncertainties should we incline, from caution, to a pessimistic view? Or should we avoid incurring expense until an unfavourable outcome looks probable?

2 Role of each industry

Coal. Does more need to be done to enable the coal industry to make its full contribution?

Electricity/Nuclear

(a) Is our present nuclear programme sufficient to create a base for the nuclear expansion which may be needed in the late 1980s and 1990s?

(b) Should we build a demonstration fast reactor either on our own or in international partnership?

(c) Should we build further fossil-fuelled power stations and should they be coal-fired, dual-fired or oil-fired?

(d) What programme of power station building (nuclear or fossil-fuelled) is necessary to maintain the health of the electrical manufacturing industry?

Gas. Do present policies provide for depletion of offshore gas reserves at an appropriate rate? Or do they risk over-rapid depletion, damaging the other industries in the process?

Oil. Present policy on depletion of North Sea oil is to encourage a rapid build up until 1980; and to seek, within the very real constraints imposed by the terms of existing production licences and the need to give producers adequate incentives, to preserve freedom to adopt a more conservative depletion profile thereafter. Is this stance adequate for the present, and towards what degree of restraint on depletion should we be working?

(Policies for depletion of both gas and oil involve balancing present benefits against the future; and an assessment of the risk of future shortfalls in supplies and of the insurance premium that it is worth paying to guard against that risk.)

Research. Is the current emphasis in research and development appropriate? Is the current emphasis on nuclear research and development appropriate? Do expenditures on energy conservation and unconventional sources match opportunities and need? (Expenditure on unconventional sources in another form of insurance against future possible inadequacy of conventional supplies. How big an insurance premium are we prepared to pay?)

SESSION 2 (2.30 pm—4.00 pm)

Agenda

The role of conservation, the consumer and the environment.

Areas for decisions

Development of the energy conservation programme.
Whether to adapt tariff structures for social ends.
Methods of payment.

Environmental constraints on development of the fast reactor.

Topics for discussion

1 Conservation

(a) Present policy is that energy conservation is to be encouraged in so far as it is economic, but not further. Is this right?

(b) In so far as methods of energy use are seen to be below the economic optimum, how positively and by what means should the Government seek to intervene to improve matters?

2 The consumer

(a) Should flat rate tariffs be introduced to benefit small consumers?

(b) Should changes be made in payment systems?

3 The environment

What further action is needed to resolve problems of radio-active waste management?

SESSION 3 (4.00 pm—5.00 pm)

Agenda

Policy and participation; how should energy policy be formulated in future? Priorities in developing resources; relationships between energy industries, and between the industries, the consumer and the Government.

Areas for decision

Future machinery for energy policy formulation.

Topics for discussion

1 Objectives and criteria

The objectives of UK energy policy have traditionally been formulated in some such terms as follows:

The energy needs of the country should be met at minimum cost in real resources over time, while, paying due regard to security of supply, to public safety, to the protection of the environment, and to the social consequences of change.

The Secretary of State said in reply to a Parliamentary Question on February 16, 1976:

'The governing assumptions and methods by which Energy policy has developed over past years now need to be re-examined in the light of changing circumstances. New criteria, for example the need for long term conservation measures and the need to develop all the fuel industries on a secure basis to meet the expected energy gap in the 1990s and beyond are required. The key role of government in developing an integrated energy policy in conjunction with the industries and those that work in them also requires further consideration. This should be done as publicly as possible so that the interests of domestic and industrial consumers long term, as well as short term can be safeguarded, along with the interests of supplying industries, and the community as a whole. I intend to proceed on this basis.'

Questions:

(a) How far do the traditional objectives still hold good? In particular:

(b) Should wider industrial, regional, economic, trade and social considerations (such as the 'Right to Warmth') be brought formally within the definition of energy policy objectives? If so which? Is it desirable to retain a clear appreciation of the cost of meeting social objectives, and would the integration of energy policy and social objectives make this more difficult?

(c) International dimension. How should these objectives be related to objectives agreed in the Community and in other international fora?

2 Relationships with industries

What systems can be devised to improve relationships within each industry in order to meet individual and national objectives? What is the proper role of planning agreements and industrial democracy and how are the interests of consumers to be represented? What improvements should we look for, what organisational structures might achieve our objective of making the best use of national resources?

3 Relationships between industries

What problems arise in seeking to associate the private sector of the oil industry in energy planning, in development of work already in progress between the Government and the nationalised energy industries?

4 The formulation of energy policy

(a) Is a new body required to conduct energy planning and follow up the Conference?

(b) How large should such a body be, and what interests should be represented? Should it be executive or advisory?

(c) What role would any such body play in relation to the UK's contribution to the formulation of international energy policy?

(d) Should the Government do more to expose its energy policy thinking for public debate, eg by issue of an energy policy Green Paper?

5 The implementation of energy policy: role of Government

(a) How positively, and in how much detail, should the Government seek to plan the energy economy? Should the Government concentrate on strategy, allowing the industries to operate within an agreed framework or should more comprehensive national planning be undertaken?

(b) To what extent is there consumer freedom of choice and how important is it?

(c) How far do the benefits flowing from competition between the energy industries outweigh the disadvantages?

(d) What is the appropriate role for prices? In particular should prices be thought of simply as instruments for bringing about a pattern of supply and demand predetermined by reference to external factors? Or should prices be determined by reference, for example, to costs and be given a role in determining what the most desirable pattern is?

(e) If an energy strategy requires an industry to take action which appears to it to be contrary to its own interests, how should the conflict be resolved? By placing it under a statutory duty, eg by the issue of directives, by adjustment of the framework (eg by subsidies) so as to make the two interests coincide, or by what other means?

(f) Can relative prices be relied on to confine high quality fuels for high quality purposes (on the assumption that high quality fuels, including electricity, will command a high price and will be used only where the consumer sees specific advantage in doing so?) Or should the Government plan to direct the purposes for which different fuels should be used?

World energy resources: present position and prospects

I INTRODUCTION

1 This paper is intended as a background document to put our specifically UK problems in their proper world wide context. It reviews past trends in world energy growth and changing patterns of supply. It then considers energy reserve estimates and likely availability of fossil fuels, nuclear power and alternative energy sources, their associated problems, and the various factors influencing future demand for energy. Finally some tentative conclusions are drawn, both for the world at large and for the UK. The uncertainty involved in forecasts should be borne firmly in mind, particularly for the period beyond 1985.

II HISTORIC TRENDS IN WORLD ENERGY SUPPLY AND DEMAND

Energy consumption 1962–1975

2 Energy consumption is closely related to GDP. The period up to the oil crisis of 1973 was one of rapid growth in both. Total world primary energy consumption between 1962 and 1974 increased by 82 per cent from some 3,300 million tons of oil equivalent (mtoe) to 6,100 mtoe. Within these totals OECD* consumption increased by 66 per cent from 2,100 to 3,400 mtoe in 1973. Consumption by the centrally planned economies (Communist countries) nearly doubled and consumption by developing countries, including OPEC* more than doubled. Between 1973 and 1975 total energy consumption in OECD countries fell slightly, as did GDP.

3 Tables 1 and 2 (pages 10 and 11) set out world energy consumption by main country groups and show the volume and proportion of total energy demand met by the different fuels. The tables show that between 1962 and 1974 oil and gas together met more than 80 per cent of the increase in energy consumption with oil alone meeting more than half the increase. Oil has become the principal marginal fuel in energy supply. However, since 1973, with the oil price explosion and the temporary reduction in energy consumption, its role has been slightly reduced. In the OECD area oil consumption in 1975 accounted for some 51 per cent of total energy consumption; natural gas and solid fuel accounted for about 20 per cent each and the remainder was provided by hydro-electricity and nuclear power. In the developing countries the proportion of oil consumption in total demand for commercial energy in 1974 was significantly higher at about 64 per cent, but this figure ignores local fuels such as wood.

4 Of OECD total energy requirements in 1975, 64 per cent were met by indigenous production. Between 1962 and 1973, although the volume of indigenous production in the OECD area increased, as a percentage of total energy requirements it declined. Since 1973 this latter trend has been reversed, but for how long remains to be seen.

Energy production

5 Table 3 (p. 12) shows production of crude oil and natural gas liquids (NGL) by area from 1960–1975. This indicates that total world oil and gas production temporarily peaked in 1973. In 1960 the OECD and OPEC areas were each producing about 39 per cent of the world's total production of crude oil and NGL, but in 1973 OPEC were producing 54 per cent compared with 23 per cent in OECD. In 1975 OPEC's share of production fell back to about 50 per cent and more than a fifth of their capacity is currently lying idle.

6 Table 4 (p. 12) gives solid fuel and natural gas consumption for the years 1962–1975. OECD consumption of solid fuel has remained fairly steady over the period, while it has grown in volume terms in the centrally planned economies and the non-OPEC developing countries. The figures for indigenous production as a percentage of consumption show that both coal and natural gas are for the most part consumed within the

* For members see page 9.

Table 6 Duration of recoverable world energy reserves (thousand million tons of oil or oil equivalent)

	Total recoverable reserves (including probable and undiscovered)	*1974 consumption*	*Duration of reserves in years at exponential consumption growth rates of:* 0%	2%	4%
Oil	233	2·8	82	49	37
Gas	171	1·1	155	72	51
Coal and Lignite					
Recovery rate of 10 per cent	645	1·7	379	108	71
Recovery rate of 50 per cent	3,225	1·7	1,895	185	110
Uranium					
Up to 30$/lb in thermal reactors	59				
Up to 30$/lb in fast reactors	2,932 (excluding CPEs)				

countries in which they are produced and that international trade in these commodities is a small proportion of total consumption.

III RESERVES

General comments on duration of reserves

7 Estimates of reserves are fraught with uncertainty. Apart from differences in the definitions and conventions used for different purposes estimates of as yet undiscovered reserves must obviously be highly speculative.

8 The conventional way of assessing reserves is to list them under 'proved', 'probable and possible', and 'undiscovered' headings. It is essential to concentrate attention on recoverable reserves, though the extent and timescale of recovery is bound to vary with future economic and technological circumstances. In the short term proved recoverable reserves are important and the ratio of current production to proved reserves is often quoted. But ultimately recoverable reserves, however uncertain, are a more valid guide to long term energy prospects.

9 Although reserves for different types of energy are assessed in different ways, and estimates differ widely, an attempt has to be made to bring them onto a common energy basis if they are to be compared. Table 5 (p. 13) gives recent published reserve figures converted into thousand million tons of oil equivalent. To set them in perspective the duration of the reserves is expressed in Table 6 in years at given exponential consumption growth rates.

10 Table 6 gives a crude illustration of the effects of exponential growth, with consumption growing to a point where it suddenly collapses to nothing with the exhaustion of total recoverable reserves. In practice, the pattern of discovery of energy reserves and the increasing cost of exploiting them as recoverable reserves become depleted would not permit such a consumption profile. Production of a particular energy source would rise to a plateau and then decline gradually, with consumers being forced either to switch to alternative sources or reduce total energy consumption. This situation is illustrated in the case of oil in the Figure on page 6.

The prospects for the individual energy sources are discussed in more detail in the following paragraphs. The prospects for energy demand as a whole are discussed in a separate section (paras. 21–25).

Oil reserves

11 More than half the world's proved oil reserves are in the Middle East, and 68 per cent are in OPEC countries as compared with only 10 per cent in OECD countries and about 1½ per cent in the UK. The Middle East's share of ultimately recoverable (including proved) reserves may be about a third and a major part of undiscovered reserves may be in the USSR. But undiscovered reserve estimates are extremely uncertain, and it is not clear how fast future exploration rates could be stepped up to find them (eg deep sea production). Future growth of oil consumption is therefore constrained not only by the limited size of ultimate reserves but also by the rate at which they can be exploited. Ultimately recoverable reserves could last for more than 80 years at current rates of consumption but for less than 40 years with four per cent per annum growth, which compares with a historic growth rate between 1962 and 1973 of nearly eight per cent. As the Figure suggests the

Theoretical growth curve—exponential growth at 4% p.a.

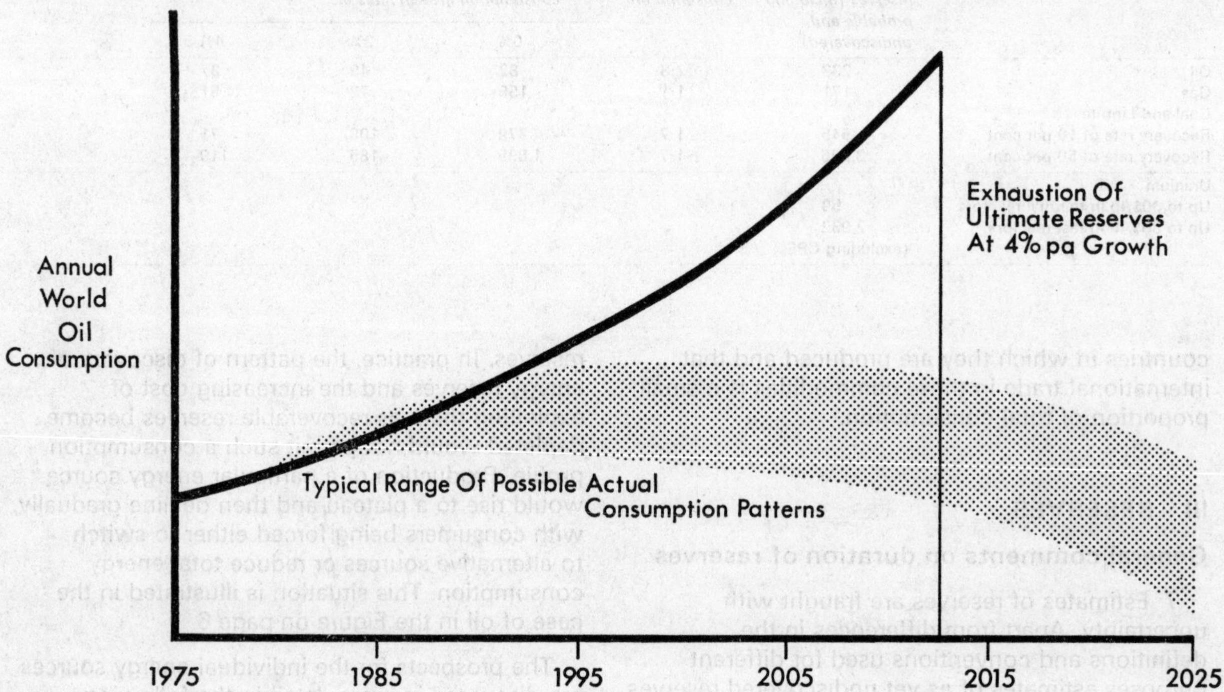

practical consequence of this situation might be declining world oil production before the turn of the century.

12 In addition to these 'conventional' reserves there are substantial oil resources locked up in tar sands and oil shales. Relatively small scale production of these resources is already under way, but the economics of exploiting them are not yet fully proved and we cannot estimate when commercial production on a sufficiently large scale to influence the world demand and supply picture might start.

Gas reserves

13 Table 6 shows that theoretically total recoverable gas reserves would, with a consumption growth rate of four per cent per annum, last about 50 years. Their geographical distribution is somewhat different from that of oil reserves, with the USSR in a very strong position. Transport costs for gas are much greater than for oil over long distances. Importing gas by sea to supplement indigenous production is costly and difficult.

Coal reserves

14 The coal reserves shown in Table 5 are sub-divided into basic categories, 'Proved recoverable' and 'Proved + Probable'. For the latter no published estimates of recoverability exist and a range of 10 per cent to 50 per cent recovery is quoted in the table. Coal reserves are obviously substantial, but there is a wide range of possible recovery due largely to the uncertainty about the future economic climate and the energy price levels against which coal would have to compete. Coal can be converted into both gas and oil though the processes are not generally economic in present circumstances. Moreover the location of reserves mainly in the USA and USSR could pose environmental and transport problems in their exploitation on a scale necessary to enable coal to take on a major part of the international energy role up to now performed by oil.

Uranium reserves

15 The uranium reserves shown in Table 5 are quoted in terms of the oil consumption at oil-fired

power stations which would produce the same amount of electricity as that obtainable from the published reserves of uranium when used in current designs of thermal reactors and possible fast reactors. There must of course be great uncertainty about the conversion efficiencies of uranium into heat, particularly for the fast reactor. There is also great uncertainty about the size of uranium reserves. Exploration has not been as extensive as for fossil fuels and we have little information about higher cost reserves which could prove economically attractive in due course. These factors, and the rapid percentage growth rate of what is still in scale an infant industry, make it unrealistic to give figures for the duration of reserves as for the other fuels in Table 6.

16 The significance of the fast reactor is that it can extract much more energy from uranium than thermal reactors. Prototypes are already operating in some countries. But there are still difficult issues to resolve and work on safety and environmental issues will be particularly important. The next stage of development will involve the construction of full-scale demonstration reactors and it is unlikely that large-scale ordering in most industrial countries would begin before the late 1980s or early 1990s.

17 The uranium reserves figures do suggest that without the fast reactor uranium availability could become a serious constraint on the rate of nuclear power station ordering before the end of the century.

Sources other than fossil and nuclear fission

18 In the long term a number of other sources of energy may be developed.

The most important are:

(a) nuclear fusion
(b) solar energy
(c) wind power
(d) wave power
(e) tidal power
(f) geothermal power

19 *Nuclear fusion* has enormous potential, but the development of a controlled thermo-nuclear fusion reactor—if it can be done at all—will take many years, and it is unlikely to make any contribution to energy supplies until well into the next century. *Solar heating and cooling* are already commercially viable in some markets, being particularly attractive in certain parts of the world such as the southern USA and Japan. But development seems likely to be slow, particularly given the modest annual rate of change in the housing stock in most countries of the world. *Photovoltaic devices* (methods whereby sunlight can be converted into electrical energy) would require large areas of land, fairly constant sunshine and a massive reduction in costs before they could be economic. *Biomass* (where plants are cultivated specifically for processing into fuel) has some limited potential, but mainly in tropical countries. *The wind, the waves and the tides* have all been considered as possible methods of electricity generation. As compared with conventional methods they all suffer to some degree from uncontrollable variability of output. However *wave power* appears worth further investigation and the UK has recently announced an expanded research programme in this area. A 240 MW *tidal power* scheme has been generating electricity at Rance in in France for a number of years but the number of suitable sites world-wide is limited and the environmental and other difficulties are very great. *Geothermal power* is also exploited for electricity, particularly in the USA, New Zealand, and Italy but only in exceptional locations. Both economics and technology elsewhere are much more dubious.

20 In any discussion of alternative energy sources it is important not to underestimate the timescales involved in development. It took 25 years of research and development into the peaceful uses of nuclear power to bring it to a point where it supplied one per cent of world demand And even if the technology is not particularly complex there are often other factors, such as the need to tie in with changes in the housing stock and resistance to change by millions of individual consumers, which will tend to slow down progress. Above all, none of the alternative sources are clearly economic at present. It seems unlikely therefore that they can supply more than at most 10 per cent of world energy by the end of the century even though their long-term importance may be greater.

IV FUTURE WORLD ENERGY DEMAND AND SUPPLY BALANCE

21 Whether or not the inevitable shift away from the world's present and immediately prospective dependence on oil to other energy sources can be achieved without disruption depends crucially on the evolution of world energy demand. The problems of forecasting are formidable. Demand will continue to depend mainly on economic growth (probably the most important factor of all), on price levels and on conservation policies. The relationship between the

different factors is difficult to determine and historically derived formulae will not necessarily give a valid guide to the future. However, detailed analysis since the 1973–74 oil price explosion may in due course offer some hope of better forecasts.

22 Up to 1973 world energy consumption was growing at the rate of about five per cent a year. Table 1 shows that in 1974 total world energy requirements rose by only 0·4 per cent while the provisional OECD figures for 1975 suggest that total consumption actually fell in the OECD area as did GDP.

23 A recent study by the International Energy Agency (IEA) submitted to the Energy Commission of the Conference on International Economic Co-operation (CIEC) gives forecasts of world energy demand and the pattern of supply up to 1985, based on a variety of assumptions. For the OECD area the IEA looked at energy requirements on four different economic growth rate assumptions varying from 2·5 per cent to 4·8 per cent per year from 1974 to 1980 and from 2·5 per cent to 4·5 per cent per year from 1980 to 1985. They also looked at variants of these reflecting possible new policies designed to intensify energy conservation and accelerate the development of new energy sources within the OECD area. In making the forecasts of supply and demand the IEA also made the working hypothesis that energy prices would remain at present levels, allowing for inflation, throughout the period. This examination suggested that for the OECD area alone total energy requirements (excluding those for marine bunkers) might lie in the range 3900–4300 mtoe in 1980 and 4400–5400 mtoe in 1985. The wide range of possible assumptions for the period beyond 1985 makes forecasting even more problematical and the IEA study submitted to the CIEC did not include any forecasts for this period.

24 The IEA study included forecasts for the non-OECD areas up to 1985, although these must be treated with caution as they are very uncertain. Table 7 gives the results of these forecasts together with the forecasts for the OECD area based on one particular set of assumptions. These assumptions are that the GDP of OECD countries will grow by some 4·3 per cent per annum from 1974 to 1980 and some four per cent from 1980 to 1985. Economic growth rates in developing countries are assumed to be somewhat higher than this and for Eastern Europe and the USSR growth of between 2½ per cent and 3 per cent per annum is assumed. Table 7 shows that total energy production in the OECD area might rise from 2270 mtoe in 1974 to about 2800 mtoe in 1980 and 3500 mtoe in 1985. Total energy requirements (including marine bunkers) in the OECD area might rise to about 4300 mtoe in 1980 and 5300 mtoe in 1985. For the OPEC area production of some 1850 mtoe in 1980 and 2150 mtoe in 1985 might be possible, leaving exports of about 1650 mtoe in 1980 and 1900 mtoe in 1985. Forecasts for other developing countries are very much more uncertain due to the lack of data but it is possible that energy production might be about 600 mtoe in 1980 and 900 mtoe in 1985 with consumption round 800 mtoe in 1980 and 1000 mtoe in 1985. In the USSR and Eastern Europe consumption might be 1700 mtoe in 1980 and 2100 mtoe in 1985 as compared with some 1300 mtoe in 1974. The forecasts suggest that total world commercial energy consumption might grow by some 4½ per cent per annum between 1974 and 1985.

25 Although the IEA figuring does not go beyond 1985 the likely further rise in energy demand does suggest that the situation could be more difficult in subsequent years. A 3½ per cent exponential growth rate for oil as in Table 7 would be a very substantial reduction on a 1962–1973 growth rate of nearly eight per cent. The latter, if continued, would have completely exhausted all ultimately recoverable reserves by the end of the century. But the reserve analysis suggests that continued exponential growth at even 3½ per cent would have to be substantially curbed well before the end of the century. The burden to be borne by other energy sources would therefore become even heavier, and the limitations and difficulties associated with these sources have already been discussed.

V FUTURE ENERGY PRICES

26 The rate of growth in demand for energy, the availability of new sources of fossil or nuclear fuel and of other sources of energy, and political factors, will all affect the future course of energy prices. The complex inter-relation of these elements makes it impossible to produce any firm estimate of future prices. Evidence since the rapid rise in oil prices suggests that the economic laws of supply and demand will continue to have some effect, so that, if there were to appear an effective surplus of energy supplies over any significant time prices would fall. Conversely, as demand increases and supplies of readily accessible energy fall short of demand it is to be expected that prices will rise, and it would not be unreasonable to expect a significant increase in energy prices compared with other prices by the end of the century.

VI CONCLUSIONS

27 The future pattern of energy supply and demand is tremendously uncertain. Just as no one more than dimly foresaw the magnitude and abruptness of the oil price explosion in 1973/1974 there could be further unforeseen developments which will throw energy plans into disarray. But the timescale of energy developments demands long-term planning, and it would seem prudent to assume a continuing rise in energy demand, even if at a somewhat lower rate than in the past. This is because a lower demand level would be inconsistent with further economic development and rising living standards.

28 The world supply problem must therefore be tackled on a broad front:

(a) Conventional fossil fuels (oil, gas, and coal) must be exploited to the full wherever it is economic to do so. In particular this means a major expansion of the coal industry world-wide. Despite the difficulties we can envisage that coal might eventually take the place of many of the uses currently performed by conventional oil. Extraction and distribution technology will need to be developed further to make this possible at reasonable cost.

(b) The future development of the electricity supply industry will require an increasingly important role to be played by nuclear power. Uncertainties about the adequacy of uranium resources mean that the fast breeder reactor will need to be developed to a point where it can be ordered commercially on a large scale in the 1990s if necessary.

(c) Research into the alternative sources of energy such as solar energy and wave power should continue both for their long-term potential and because of the help that some of them may be able to provide in meeting energy demand by the end of the century.

(d) Conservation measures will continue to play an important role, but they cannot prevent all future energy growth without harming economic growth.

29 For the UK the conclusion must be that we must not allow ourselves to become complacent about the present strength of our energy resources. We too need to press ahead on a broad front, bearing in mind that when our oil and gas supplies begin to decline in the 1990s or thereabouts we must not find ourselves once more unnecessarily dependent on large and increasing imports of energy at possibly even higher prices than now.

OECD MEMBERS

Australia
Austria
Belgium
Canada
Denmark
Finland
France
Germany
Greece
Iceland
Ireland
Italy
Japan
Luxembourg
Netherlands
New Zealand
Norway
Portugal
Spain
Sweden
Switzerland
Turkey
United Kingdom
United States

OPEC MEMBERS

Algeria
Ecuador
Gabon
Indonesia
Iran
Iraq
Kuwait
Libya
Nigeria
Qatar
Saudi Arabia
United Arab Emirates
Venezuela

Table 1 World energy consumption (volume)

	Year	Oil(1)	Natural Gas	Solid Fuels	Primary Electricity: Hydro	Primary Electricity: Nuclear	Total (TER) (2)	TER as per cent of world total (3)	Indigenous Production (2)	Net (2) Imports
		Million tons oil equivalent							Mtoe	
		1	2	3	4	5	6	7	8	9
1 OECD TOTAL	1962	869·7	350·2	673·2	160·7	1·8	2,055·6	61·6	1,624·6	438·8
	1968	1,362·5	513·5	694·1	191·2	11·9	2,773·2	61·1	1,967·4	832·7
	1972	1,823·4	686·8	647·0	217·0	33·6	3,407·8	60·2	2,239·0	1,190·3
	1973	1,952·1	705·1	686·9	222·0	45·5	3,611.6	59·8	2,271·1	1,347·0
	1974	1,847·9	707·0	680·1	240·0	57·3	3,532.3	58·3	2,258·7	1,302·8
	1975(4)	1,757·0	667·0	667·0	246·0	76·0	3,413·0	n/a	2,186·0	1,227·0
2 CENTRALLY PLANNED ECONOMIES (CPE) (including Yugoslavia)	1962	179·2	82·0	657·6	32·0		950·8	28·5	1,001·5	—50·7
	1968	295·6	186·0	756·3	47·4		1,285·3	28·3	1,361·7	—76·4
	1972	423·7	260·0	882·2	59·9		1,625·8	28·7	1,679·7	—53·9
	1973	481·2	278·4	910·1	60·3		1,730·0	28·6	1,780·0	—50·0
	1974	517·2	299·3	917·1	66·6		1,800·2	29·7	1,868·5	—68·3
	1975									
3 OPEC	1962	46·1	10·6	1·1	0·3		58·1	1·7	556·4	—498·3
	1968	59·5	19·2	0·7	1·2		80·6	1·8	992·3	—911·7
	1972	67·3	33·8	1·4	4·3		106·8	1·9	1,449·6	—1,342·8
	1973	84·4	39·8	1·5	5·0		130·7	2·2	1,656·7	—1,526·0
	1974	97·3	43·4	1·5	5·0		147·2	2·4	1,651·4	—1,504·2
	1975									
4 OTHER DEVELOPING COUNTRIES	1962	141·6	11·2	63·0	19·7		235·5	7·1	162·5	73·0
	1968	217·1	21·0	75·9	33·4		347·4	7·6	249·6	97·8
	1972	294·5	33·3	80·6	46·6		455·0	8·0	312·3	142·7
	1973	331·7	38·0	83·1	50·9		503·7	8·3	329·5	174·2
	1974	332·6	40·2	91·0	56·9		520·7	8·6	355·7	168·5
	1975									
5 TOTAL DEVELOPING COUNTRIES (3+4)	1962	187·7	21·8	64·1	20·0		293·6	8·8	718·9	—425·3
	1968	276·6	40·2	76·6	34·6		428·0	9·4	1,241·9	—813·9
	1972	361·8	67·1	82·0	50·9		561·8	9·9	1,761·9	—1,200·1
	1973	416·1	77·8	84·6	55·9		634·4	10·5	1,986·2	—1,351·8
	1974	429·9	83·6	92·5	61·9		667·9	11·0	2,003·6	—1,335·6
	1975									
6 TOTAL (5) WORLD	1962	1,243·5	454·0	1,422·7	214·5		3,334·7	100·0		
	1968	1,949·2	739·8	1,562·4	285·1		4,536·5			
	1972	2,629·9	1,014·0	1,651·2	361·6		5,656·7			
	1973	2,870·9	1,060·7	1,724·1	383·7		6,039·4			
	1974	2,815·8	1,090·0	1,734·3	426·1		6,066·2			
	1975									

(1) Includes bunkers and non-energy use.

(2) For OECD countries, Indigenous Production and Net Imports do not add to Total Energy Requirements because of stock changes; for other countries, stock changes are included in the source figures used in calculating Total Energy Requirements, Indigenous Production and Net Imports.

(3) Because of the differences in reflecting stock changes (see Note (2) above) and other differences between the two sources used, percentages of World Total are approximate.

(4) Preliminary: subject to revision when final figures are available.

(5) Includes other countries not included in above main country groups.

Source: see Table 2

Table 2 World energy consumption (percentage share)

	Year	% Share in Total: Oil[1]	Natural Gas	Solid Fuels	Primary Electricity: Hydro	Nuclear	Total (TER) Mtoe[2]	TER as[1] per cent of world total[3]	% Share in Total: Indigenous Production[2]	Net Imports[2]
		1	2	3	4	5	6	7	8	9
1 OECD TOTAL	1962	42·3	17·0	32·7	7·8	0·1	2,055·6	61·6	79·0	21·3
	1968	49·1	18·5	25·0	6·9	0·4	2,773·2	61·1	70·9	30·0
	1972	53·5	20·1	19·0	6·4	1·0	3,407·8	60·2	65·7	34·9
	1973	54·1	19·5	19·0	6·1	1·3	3,611·6	59·8	62·9	37·3
	1974	52·3	20·0	19·3	6·8	1·6	3,532·3	58·3	63·9	36·9
	1975[4]	51·5	19·5	19·5	7·2	2·2	3,413·0	n/a	64·0	36·0
2 CENTRALLY PLANNED ECONOMIES (CPE) (including Yugoslavia)	1962	18·8	8·6	69·2	3·4		950·8	28·5	105·3	—5·3
	1968	23·0	14·5	58·8	3·7		1,285·3	28·3	106·0	—6·0
	1972	26·1	16·0	54·3	3·7		1,625·8	28·7	103·3	—3·3
	1973	27·8	16·1	52·6	3.5		1,730·0	28·6	102·9	—2·9
	1974	28·7	16·6	50·9	3·7		1,800·2	29·7	103·8	—3·8
	1975									
3 OPEC	1962	79·3	18·2	1·9	0·5		58·1	1·7	957·7	857·7
	1968	73·8	23·8	0·9	1·5		80·6	1·8	1,231·1	—1,131·1
	1972	63·0	31·6	1·3	4·0		106·8	1·9	1,357·3	—1,257·3
	1973	64·6	30·4	1·1	3·8		130·7	2·2	1,267·6	—1,167·6
	1974	66·1	29·5	1·0	3·4		147·2	2·4	1,146·8	—1,046·8
	1975									
4 OTHER DEVELOPING COUNTRIES	1962	60·1	4·8	26·7	8·4		235·5	7·1	69·0	31·0
	1968	62·5	6·0	21·8	9·6		347·4	7·6	71·8	28·2
	1972	64·7	7·3	17·7	10·2		455·0	8·0	68·6	31·4
	1973	65·9	7·5	16·5	10·1		503·7	8·3	65·4	34·6
	1974	63·9	7·7	17·5	10·9		520·7	8·6	67·9	32·1
	1975									
5 TOTAL DEVELOPING COUNTRIES (3 +4)	1962	63·9	7·4	21·8	6·8		293·6	8·8	244·8	—144·8
	1968	64·6	9·4	17·9	8·1		428·0	9·4	290·2	—190·2
	1972	64·4	11·9	14·6	9·1		561·8	9·9	313·6	—213·6
	1973	65·6	12·3	13·3	8·8		634·4	10·5	313·1	—213·1
	1974	64·4	12·5	13·8	9·3		667·9	11·0	300·0	—200·0
	1975									
6 TOTAL[5] WORLD	1962	37·3	13·6	42·7	6·4		3,334·7	100·0		
	1968	43·0	16·3	34·4	6·3		4,536·5	↓		
	1972	46·5	17·9	29·2	6·4		5,656·7			
	1973	47·5	17·6	28·5	6·4		6,039·4			
	1974	46·4	18·0	28·6	7·0		6,066.2			
	1975									

Footnotes: see Table 1

Source: OECD countries—OECD Statistics of Energy; all non-OECD countries—United Nations Series J, No. 18 and UN. Submission to the CIEC (Tables 1 and 2), except that

(i) tons of coal equivalent have been converted to tons of oil equivalent by a factor of 0·7;

(ii) hydro and nuclear electricity have been multiplied by 3·33 to reflect an assumed generating efficiency rate of 30 per cent.

The difference between 'total supply' and 'total consumption' is assumed to be non-energy consumption of petroleum products and refinery fuels (see paragraph 2, page ix, Series J, No. 18) and is added to oil consumption.

Table 3 Production of crude oil and natural gas liquids (millions tons)

	1960		1962		1968		1972		1973		1974		1975	
		% World Total		% World Total		% World Total		% World Total		% World Total		% World Total		% World Total
1 OECD	424·9	39·1	454·2	36·0	584·3	29·3	654·7	24·9	653·5	22·8	628·1	22·0	599·9	22·1
2 CPE	165·1	15·2	203·0	16·1	344·9	17·3	452·9	17·3	502·3	17·5	542·2	19·0	586·4	21·6
3 OPEC	431·6	39·7	530·0	42·0	940·3	47·2	1,364·3	51·9	1,551·0	54·0	1,530·4	53·0	1,363·5	50·2
4 Other Developing	65·1	6·0	75·4	5·9	119·7	6·0	152·1	5·8	156·1	5·4	152·9	5·3	163·2	6·0
5 Total Developing (3+4)	496·7	45·7	605·4	47·9	1,060·0	53·2	1,516·4	57·7	1,707·1	59·5	1,683·3	58·9	1,524·7	56·1
6 Total World (1+2+3+4)	1,086·8	100	1,262·8	100	1,991·3	100	2,630·3	100	2,868·7	100	2,858·6	100	2,716·8	100

Source: OECD Countries — OECD Statistics of Energy;
Rest of the World — 1960 – 1972 UN Series J, No. 15: Table No. 8
1973 – 1975 Oil Industry Source

Table 4 Solid fuel and natural gas balances

		Solid Fuels		Natural Gas (1)	
		Consumption (mtoe)	Indigenous Production as % of Consumption	Consumption (mtoe)	Indigenous Production as % of Consumption
1 TOTAL OECD	1962	673·2	97·2	350·2	99·6
	1968	694·1	96·8	513·5	99·5
	1972	647·0	96·1	686·8	99·3
	1973	686·9	96·1	705·1	98·9
	1974	680·1	94·7	707·0	98·3
	1975 (2)	667·0	n.a.	667·0	n.a.
2 CENTRALLY PLANNED ECONOMIES (CPE) (incl. Yugoslavia)	1962	657·6	102·6	82·0	100·0
	1968	756·3	102·6	186·0	99·6
	1972	882·2	102·3	260·0	96·6
	1973	910·1	102·5	278·4	96·8
	1974	917·1	103·0	299·3	98·1
3 OPEC	1962	1·1	90·9	10·6	100·0
	1968	0·7	86·0	19·2	106·2
	1972	1·4	78·6	33·8	134·9
	1973	1·5	73·3	39·8	133·9
	1974	1·5	80·0	43·4	131·6
4 OTHER DEVELOPNG COUNTRIES	1962	63·0	96·2	11·2	109·8
	1968	75·9	95·1	21·0	111·9
	1972	80·6	95·8	33·3	107·8
	1973	83·1	95·7	38·0	109·5
	1974	91·0	94·8	40·3	113·7
5 TOTAL DEVELOPING COUNTRIES (3+4)	1962	64·1	96·1	21·8	105·0
	1968	76·6	95·0	40·2	109·2
	1972	82·0	95·5	67·1	121·5
	1973	84·6	95·3	77·8	122·0
	1974	92·5	94·6	83·6	123·0

(1) Includes liquified natural gas; does not include natural gas liquids.

(2) Preliminary; subject to revision when final figures are available.

Source: OECD countries – OECD Statistics of Energy;
Rest of the world – United Nations Series J No. 18, and UN

Table 5 Estimated recoverable world energy reserves (thousand million tons of oil equivalent)

				Western Europe	North America	Middle East	CPEs	Other	Total World
OIL (4)	Proved (2)			2·8 (1)	5·5	50	14	17·7	90
	of which Offshore (3)			2·7	1·1	10	n.a	7·5	n.a
	Proved + Probable			3·3	8	59	18	18·7	107
	Proved + Probable + Undiscovered			11·5	29	80	59	53·5	233
GAS (6)	Proved (5)			4·2	6·6	13	16	11·2	51
	of which Offshore (3)			1·6	1·4	5·5	n.a	2·8	n.a
	Proved + Probable			4·4	10·6	15	16	14	60
	Proved + Probable + Undiscovered			9·3	28	35	55	43·7	171
High grade COAL (7)	Proved recoverable			25	112	—	126	33	296
	Proved + Probable total assuming recoverability % of	10%		28	163	—	347	18	556
		50%		140	815	—	1,735	90	2,780
Low grade	Proved recoverable			9	8	—	32	6	55
	Proved + Probable total assuming recoverability % of	10%		3	22	—	61	3	89
		50%		15	110	—	305	15	445
URANIUM (8) Up to $15/lb	Reasonably assured		Thermal reactors	1	8	—	n.a	9	n.a
			Fast reactors	51	390	—	n.a	466	n.a
	Reasonably assured + Estimated additional		Thermal reactors	2	22	—	n.a	11	n.a
			Fast reactors	81	1,082	—	n.a	584	n.a
Up to $30/lb	Reasonably assured		Thermal reactors	8	10	—	n.a	12	n.a
			Fast reactors	409	521	—	n.a	590	n.a
	Reasonably assured + Estimated additional		Thermal reactors	9	31	—	n.a	19	n.a
			Fast reactors	455	1,555	—	n.a	922	n.a

Sources:
(1) Western Europe proved oil reserves – *World Oil* August 1975
(2) Proved oil reserves excluding Western Europe – *Oil and Gas Journal* Dec 1975
(3) Oil and gas reserves – *Offshore* June 1975
(4) Oil reserves excluding proved – Moody, 9th World Petroleum Congress May 1975
(5) Proved gas reserves – *Oil and Gas Journal* Dec 1975
(6) Gas reserves excluding proved – Moody and Geiger, *MIT Technology Review* March/April 1975
(7) Coal reserves – WEC 1974
(8) Uranium reserves – *Uranium Resources Production and Demand* Dec 1975

Table 7 IEA forecasts of energy production and consumption in 1980 and 1985 (million tons of oil equivalent)

		1974		1980		1985	
		Production	*Consumption*	*Production*	*Consumption*	*Production*	*Consumption*
1							
	Solid fuels	644	680	775	819	938	987
OECD	Oil	634	1,848	871	2,185	963	2,574
	Gas	695	707	704	777	767	903
	Nuclear	57	57	219	219	474	474
	Hydro/Geo-thermal	240	240	290	290	335	335
	TOTAL	2,270	3,532	2,805	4,290	3,477	5,273
2							
USSR	Solid fuels		589		722		870
AND	Oil	472	431	600	575	645	665
EASTERN	Gas		283		375		485
EUROPE	Nuclear/ hydro	26	26	45	45	111	111
	TOTAL		1,329		1,717		2,131
3							
	Solid fuels	1	2	5	2	5	2
OPEC	Oil	1,577	97	1,705	140	1,970	185
	Gas	57	43	124	44	151	61
	Nuclear/ hydro	5	5	10	10	20	20
	TOTAL	1,640	147	1,844	196	2,146	268
4							
	Solid fuels	86	91	120	130	158	173
OTHER	Oil	170	333	263	457	400	569
DEVELOPING	Gas	46	40	77	67	118	100
COUNTRIES	Hydro/ Nuclear	57	57	110	110	190	190
	TOTAL	359	521	570	764	866	1,032
5							
	Solid fuels		1,362		1,673		2,032
	Oil	2,853	2,709	3,385	3,357	3,978	3,993
TOTAL (1)	Gas		1,073		1,263		1,549
	Nuclear/ Hydro/Geo-thermal	385	385	674	674	1,130	1,130
	TOTAL		5,529		6,967		8,704

Source: IEA submission to the 2nd session of the CIEC Energy Commission
Statistical Data and Projections. World Energy Consumption and Supply

(1) This is slightly less than the world total as no forecasts are available for centrally planned economies other than the USSR and Eastern Europe

UK energy supply and demand prospects

I INTRODUCTION

1 This paper sets out to examine on a specified range of assumptions some of the prospects and problems which could arise in meeting the UK's demand for energy over the next 15 years, with a broader look at the situation up to the end of the century.

2 Forecasting is always hazardous but it is doubly so over such a long period and in circumstances where the world economy is emerging from the grip of a major recession and is still adjusting to the energy crisis of 1973/74 and the escalation in oil prices which followed. The present paper, however, attempts to explore the implications of a range of alternative possibilities rather than attempt to predict which one of them might be the most likely; but, of course, even the situations which have been examined are not the only possible outcomes which the future may hold.

3 Two critical elements in the determination of UK energy demand and supply are the rate of economic growth and the future of world oil prices. By combining the possibility of different rates of growth with a range of energy prices it is possible to define an area of uncertainty within which many of the practical issues of energy strategy may realistically be considered.

II UK ENERGY CONSUMPTION: THE PAST AND THE PRESENT

4 The starting point for any view on the future must be a consideration of where we stand today, and this section sets out some of the main features in the current pattern of fuel production and consumption.

5 Energy consumption can be measured in three ways:

(a) the input of *primary fuels* needed to produce the energy;

(b) the *heat supplied* by the primary and secondary fuel industries after allowing for conversion losses; and

(c) the *useful energy* remaining after allowing for losses in appliances.

In 1975 the three measures of consumption were as follows:

	Million tons of coal equivalent (mtce)	*Billion Therms*
Primary fuels	320·5	80·6
Heat supplied	221·9	55·8
Useful energy	115·3	29·0

6 Over the past quarter of a century UK consumption of useful energy has grown at an average annual rate of a little under two per cent and Table 1 and Diagrams 2 and 3 (pp. 20 and 21) show how the mix of fuels to meet that consumption has changed.

7 It may be seen from these figures how the UK energy economy has evolved from a two-fuel system, dominated by coal, to a four-fuel system with oil and coal meeting two-thirds of our primary fuel needs but with natural gas and nuclear providing a growing share. The balance between the four fuels is even more marked in terms of heat supplied where the share of natural gas rises to between a fifth and a quarter of the total. These changes have occurred over the past 25 years; we are now looking forward over a period of similar length.

III FORECASTING ASSUMPTIONS

Economic growth

8 The main determinant of total energy demand is the level of economic activity. Since the 1950s an average annual growth in GDP of about 2·7 per cent has been associated with an annual growth of primary fuel consumption of 1·7 per cent—a crude energy coefficient of about 0·7. Energy consumption has also followed a cyclical pattern, for the most part closely following that of the economy as a whole.

Table 1 UK energy consumption 1950-1975

	1950	1960	1970	1975
Primary fuels input basis (mtce)				
Coal	202·6	196·7	154·4	120·3
Natural Gas	—	0·1	17·6	54·5
Oil	22·2	65·5	145·6	132·9
Nuclear and Hydro	0·9	2·6	12·0	12·8
Total	225·7	264·9	329·6	320·5

These primary fuels reach consumers in the form of heat supplied either directly or by way of the conversion industries:

	1950	1960	1970	1975
Heat supplied basis (bn Therms)				
Solid fuel[1]	36·1	31·1	17·8	10·0
Gas[3]	2·7	3·3	6·2	12·7
Oil[4]	5·6	12·8	27·3	25·8
Electricity	1·5	3·4	6·6	7·4
Total	45·9	50·6	57·9	55·9

This supply of heat has been used in the following sectors:

	1950	1960	1970	1975
Industry				
Solid fuel	15·7	13·9	8·9	5·0
Gas[3]	0·9	1·5	1·9	5·5
Oil[4]	1·7	4·5	11·4	8·8
Electricity	0·7	1·5	2·5	2·6
Total	18·9	21·4	24·7	21·9
Transport				
Solid fuel	4·5	2·8	0·1	—
Oil[4]	3·2	5·9	11·0	12·2
Electricity	0·1	0·1	0·1	0·1
Total	7·8	8·8	11·2	12·3
Domestic				
Solid fuel	11·8	11·3	7·1	4·3
Gas[2]	1·4	1·3	3·6	5·9
Oil	0·2	0·7	1·3	1·4
Electricity	0·5	1·2	2·6	3·1
Total	13·9	14·5	14·6	14·7
Other				
Solid fuel	4·1	3·1	1·7	0·7
Gas[2]	0·4	0·5	0·7	1·3
Oil	0·6	1·7	3·6	3·4
Electricity	0·2	0·6	1·4	1·6
Total	5·3	5·9	7·4	7·0

1 Coal, coke, manufactured fuel

2 Town Gas and Natural Gas

3 Town Gas, Natural Gas, and Coke Oven Gas and Gas used for non energy purposes

4 Oil and Creosote Pitch Mixtures

9 For the purposes of these forecasts two alternative trend rates of growth of GDP, 2·5 and 3·0 per cent (taking 1974 as the base), have been used; they straddle the UK's historic rate of growth and provide a sensitivity of the forecasts to changes in national income.

10 In addition to making assumptions about the growth of GDP as whole it is necessary to take a view about the associated growth of consumers' expenditure and industrial production because of their effect on energy demand and the fuel mix in different sectors of the economy. Although the growth of consumers' expenditure is likely to be constrained over the period to 1980, after that the historical relationship to GDP is assumed to re-establish itself and the estimated growth of manufacturing production is also based on the assumption that the historical relationship with GDP will be maintained in the longer term.

11 The relationship between economic activity and energy consumption is of course not constant. In particular higher prices and the advance of technology are expected on balance to produce changes leading to the more efficient use of fuel. It is in this connection that the distinction between the three measures of energy consumption is important, for according to the efficiency with which they are used in conversion processes or final use the same amounts of primary fuel or heat supplied can yield very different levels of useful energy, and changes in these relationships need to be built in to any forecasting system. The forces leading to more efficient use can be strengthened by vigorous fuel saving policies, which can have a major impact on the total energy situation and which must therefore be seen as an integral part of our energy policies, along with those for developing new sources of supply.

Energy prices

12 From a forecasting point of view, the relationship between energy demand and price has always been uncertain but there are two aspects of the price problem: first, the general level of energy prices affects the *total* level of demand, and second, the *relative* prices of individual fuels may be of considerable importance in affecting the mix of fuels used to meet a particular total energy requirement. Several factors such as the physical availability of fuels and lags in market responses influence the energy market and the share of the total taken by individual fuels at any particular time.

13 Oil price increases over the past three years put oil product prices clear of the average costs of competing primary fuels, although the gap has narrowed and there is some overlap at the margin. However, substitution between fuels within the energy sector has so far been limited to increased gas consumption in the domestic market and in industry, amd more coal used in power stations. If oil prices were to fall it would mainly affect the demand for coal, but would have a much smaller impact on the sales of gas already contracted for and on the rate of nuclear installation. Continuing

Diagram 1

Growth of GDP and primary energy consumption 1950–1975

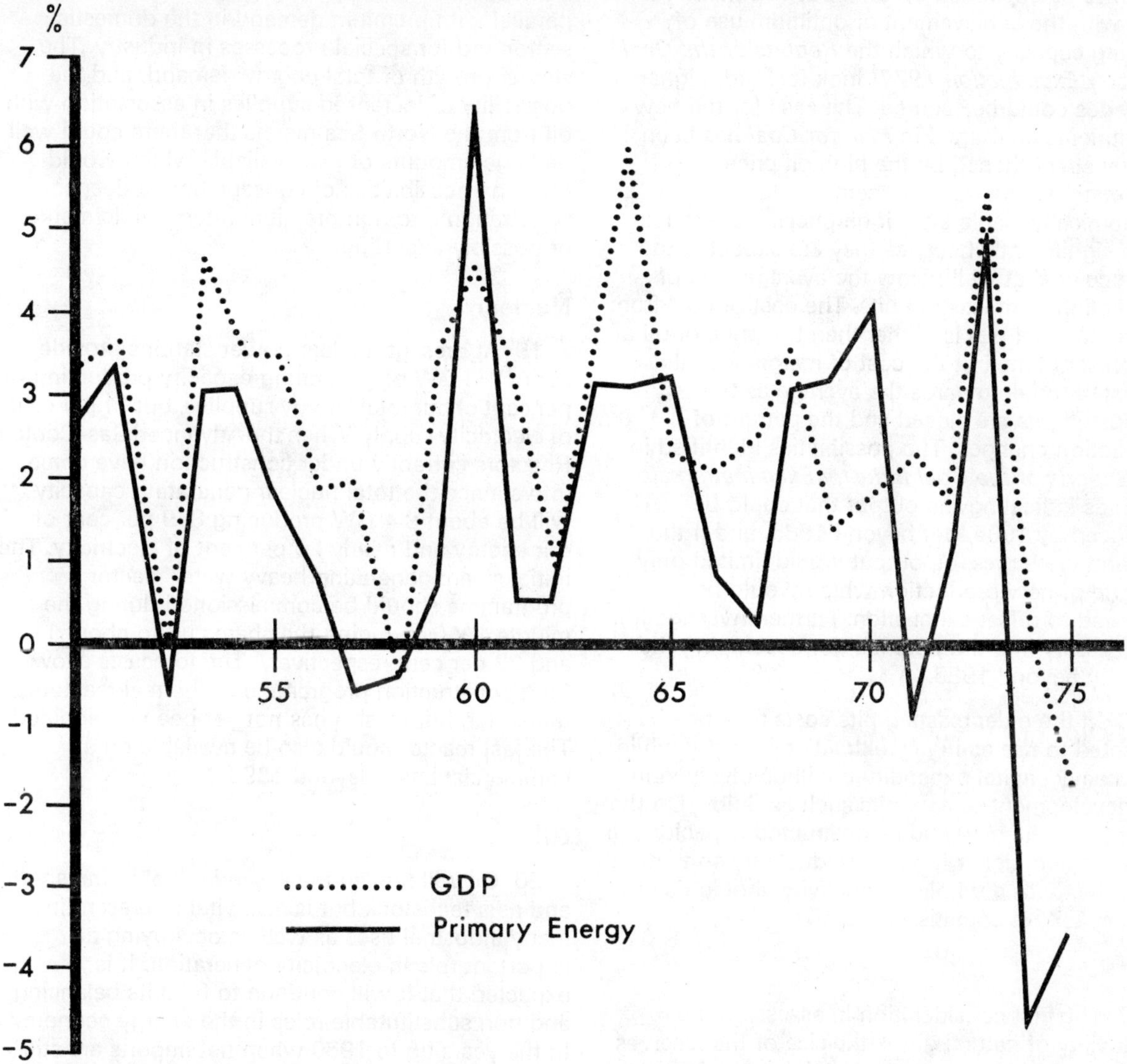

high oil prices also stimulate the development of additional supplies of other primary fuels, although there will be little effect in the short term.

14 Clearly, predicting the future price of oil is particularly difficult especially when, as at present, prices substantially exceed the costs of production. In the longer term and towards the end of the century prices are likely to rise but the path between now and then must be subject to a very wide margin of uncertainty and there is always the possibility of fluctuations occurring, taking prices, at least temporarily, well outside any trend range. In practice it is better for most forecasting purposes to concentrate on the relative prices of fuels rather than attempt to predict absolute levels and in one price situation it might be assumed that oil would be more expensive than coal and the other fuels while another price scenario would reverse this relationship.

IV FUEL SUPPLY

Coal

15 Coal output would have to decline as existing pits became exhausted but the National Coal Board (NCB) *Plan for Coal* envisages this lost

capacity being replaced by investment in new and improved capacity yielding 42 million tons of new production. This will enable coal output, up to 1985, to be expanded up to about 135 million tons; and, with the achievement of optimum use of existing capacity to which the *Reports of the Coal Industry Examination 1974** look forward, higher tonnages could be secured. The case for the new investments envisaged in *Plan for Coal* has been greatly strengthened by the high oil price expectations, but most of them would be economically viable even if oil prices were to fall quite significantly because they are expected to produce coal at well below the average cost of production from existing pits. The cost of coal from some marginal pits is higher than the price of oil at the present time but the cost of marginal coal is expected to fall towards the average as the exhausted pits are closed and the pattern of production changes. The possibilities identified in the *Reports of the Coal Industry Examination* are taken as indicating the output that could be produced by 1985, but beyond 1985, and if the demand were present, output would limited only by the rate of new production which would be achieved to offset exhaustion. Further investigation is necessary into the potential for developing the industry beyond 1985.

16 In the older existing pits costs may be expected to rise as in any extractive industry while very heavy capital expenditure will be involved in the development of new pits such as Selby. On the other hand the new and reconstructed capacity will have a much higher level of productivity so that for the industry as a whole productivity should rise as this capacity is commissioned.

Gas

17 The first consideration in assessing the availability of natural gas is the size of the reserves. The latest estimates put the remaining total on the UK Continental Shelf at the end of 1975 as about 29 trillion cubic feet (tcf) of proven reserves with a further 22 tcf as probable or possible in known discoveries. This is a cautious estimate of the total of gas which could be available to the UK for it does not include possible imports from the Norwegian sector or the possibly quite large amounts still to be discovered, probably mainly in association with oil.

18 The general view which has been taken about the use of gas has been that it should be absorbed at a rapid rate up to about 1980 mainly in order to displace costly imported oil, but thereafter a somewhat more conservationist approach should be adopted with supplies growing broadly in parallel with premium demand in the domestic sector and for special processes in industry. The slower growth of total energy demand, and the possibility of increased supplies in association with oil from the North Sea means that there could well be large amounts of gas available which would offer the possibilities of conservation, a deeper penetration into non premium interruptible markets, or possibly exporting.

Nuclear

19 At present nuclear power stations provide about 4·3 GW of generating capacity producting 4 per cent of our total energy supplies, but 11 per cent of electricity supply. When the Advanced Gas-Cooled Reactors currently under construction have come into service the total nuclear generating capacity will be about 9·4 GW producing 8–9 per cent of our energy and nearly 20 per cent of electricity. The initial steam generating heavy water reactor programme should be commissioned during the middle 1980s bringing the shares up to about 11 and 22 per cent respectively. The forecasts allow for a construction programme of thermal reactors thereafter, but its size has not yet been determined. The fast reactor could also be available on a commercial basis beyond 1990.

Oil

20 Oil will remain the essential fuel in transport and as a feedstock but is also vital at present in many industrial uses as well as occupying an important role in electricity generation. It is expected that it will continue to fulfil its balancing and non substitutable roles in the energy economy. In the years up to 1980 when net imports are still required, it is assumed that marginal requirements will always be available at a cost, although the actual level of oil prices, unless a major shift were to occur, is probably less important than whether they are higher or lower than those of alternative fuels. It is also assumed that indigenous oil from the North Sea will be priced at its full opportunity cost on the basis of world prices.

21 The availability of oil (and gas) from the North Sea has transformed the outlook for the UK economy in the medium term by removing a major debit item from the balance of payments. The estimated availability of North Sea oil by 1980 will exceed present forecasts of oil consumption and it

*Available from the Department of Energy Library, Thames House South Millbank, London SW1P 4QJ.

is very likely that this will continue through the 1980s and perhaps beyond although levels of production in this later period will increasingly depend on new developments and discoveries as well as on the Government's depletion strategy.

Electricity

Demand

22 The combined effect of recent price rises, the recession and the increase in indigenous supplies of other fuels, particularly gas, could squeeze electricity consumption severely and it is estimated that this could produce a fall in electricity sales to the domestic market up to 1980, although they are expected to recover later. Over the whole period total electricity demand, however, is still expected to grow quite rapidly, albeit more slowly than in recent years. Recovery in the late 1980s, especially if the expansion in the availability of gas were to taper off, could be rapid. The average rate of growth could be in the range of 2–3 per cent per annum up to 1980 and 3–4 per cent per annum over the whole forecast period. Total sales which were about 219 TWh in 1975 could rise to about 260 TWh in 1980 and over 400 TWh by 1990.

Generation

23 For the next four or five years the average rate of plant ordering is likely to be low as demand remains depressed, but construction is likely to pick up sharply in the early 1980s. The uncertainty on the power station ordering programme is very wide and the full range of new installed capacity up to 1990 could lie between about 15 and 30 GW. The choice of firing will depend on the relative price of fuels and the capital costs of the plant, along with such factors as flexibility and security of supply, but in general the construction of fossil fired plant is likely to be more sensitive to the economic growth (and hence total demand) prospects than to relative fuel prices. New fossil fired capacity will be needed and this is expected to be coal unless there were to be a combination of high demand together with relatively cheap oil. At this stage in the construction programme the amount of nuclear capacity is not so likely to be affected by changes in economic and oil price considerations.

New energy sources

24 In the period up to 1990, ie about 15 years ahead, it is not likely that energy supplies from the wind, waves, sun or geothermal sources will move far beyond the experimental and prototype stages. They will therefore be unable to provide more than a very small part of our total requirements until the closing years of the century. To the extent that this prognostication proves wrong and technical developments do make energy from new sources both available and less costly than that from conventional sources, then the effect would be to substitute them for those forms of energy which are more expensive or which can be conserved.

V FORECASTS OF UK ENERGY CONSUMPTION TO 1990

25 Putting all the foregoing demand and supply considerations together, Table 2 summarises in broad outlines the way in which the UK energy economy might evolve between now and 1990. Diagrams 2 and 3 are illustrative only but the values depicted lie within the ranges postulated in the table.

Table 2 UK primary fuels: supply and demand 1975-1990 (million tons of coal equivalent)

	1975	*1980*	*1985*	*1990*
Energy Uses				
Coal	120	120–125	125–135	115–150
Gas*	50	65–75	65–95	65–96
Nuclear and Hydro	13	25	30	40–55
Oil	133	100–125	100–165	115–225
TOTAL ENERGY	316	325–345	350–400	395–465
Non Energy Oil and Gas	24	35–40	40–45	50–55
TOTAL PRIMARY FUELS	340	360–385	395–445	440–520
INDIGENOUS PRODUCTION	191	375–415	390–505	390–555
of which North Sea Oil	1	160–190	165–250	165–250

*Excluding non energy uses.

VI BEYOND 1990

26 The foregoing figuring and analysis covers a period of only about fifteen years from the present day. It is thus heavily conditioned by existing patterns of production and consumption and by fairly well established past relationships between energy and other economic and technological considerations. Analysis related to the projection of past trends can provide a reliable framework for the consideration of energy policy for only a limited period ahead, and in the longer term the patterns of the past become less of an influence, their relationships weaken and the wedge of uncertainty widens.

27 As a starting point, however, we may consider what would be involved if the projected

Diagram 2

UK primary energy 1950–1990 (million tons coal equivalent)

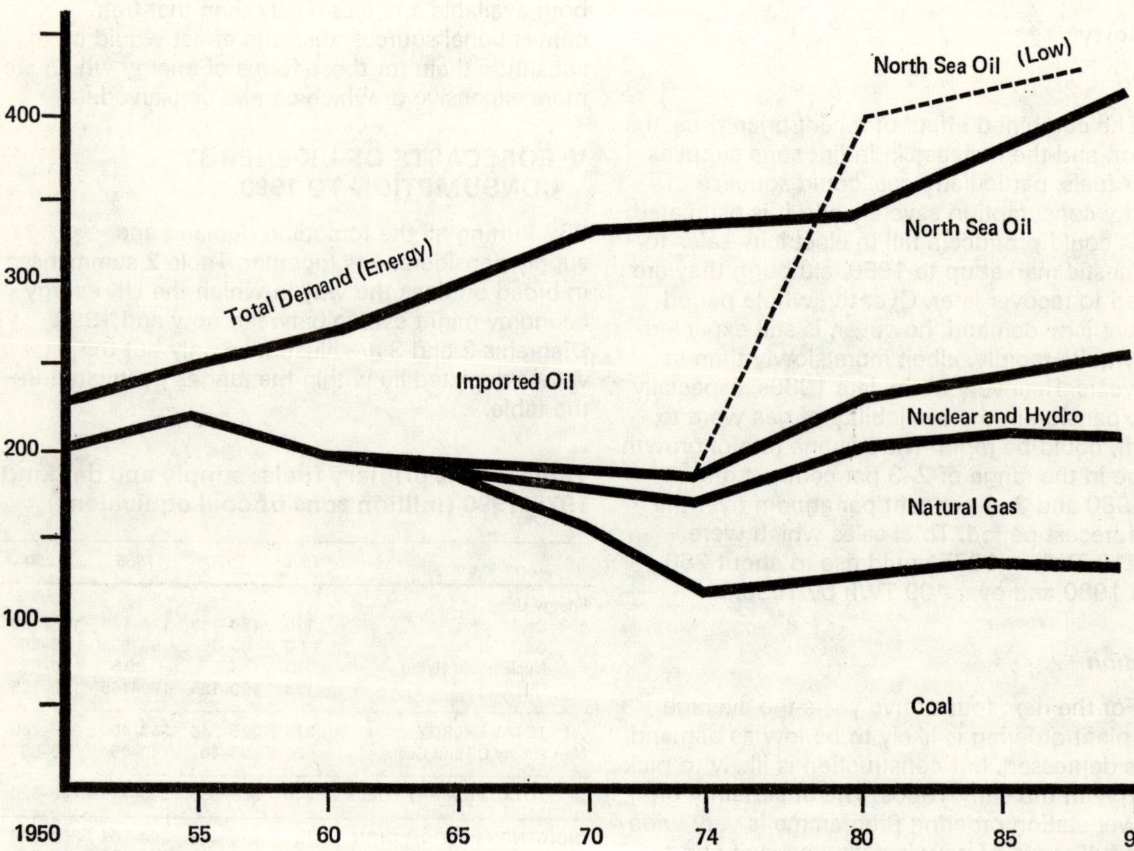

trends of the 1980s were to persist to the end of the century. With a 2 per cent annual growth rate, primary fuel consumption, which now stands at around 320 mtce in energy uses and could reach 400–450 mtce in 1990, would rise to about 550 mtce by 2000, and because within the total electricity would be growing more rapidly its share would rise from about one-third of total consumption (in primary fuel input terms) at present to over one-half by the year 2000.

28 But is this a plausible picture? It is possible that it represents an upper limit, and that under the combined effects of higher prices, market saturation, and fuel saving policies demand for energy will not continue to grow at its past rate. Higher prices would be a reflection of the fact that increasing demands would be increasingly difficult to meet, and this must tend to affect the level of our living standards and the rate at which they can improve. This effect can only be offset to the extent that vigorous conservation policies can lead to the achievement of given standards of efficiency, comfort and convenience with lowered energy inputs; since lead times are frequently long, the extent of the improvement will depend on the foundations laid in the next few years. Both the need for energy conservation, and the economic scope for it, can be expected to increase progressively as the century advances.

29 Whatever the relationship between economic growth and energy may be, and more detailed consideration of them is required, the UK cannot be isolated from the rest of the world economy. If demand, economic activity and trade are continuing

Diagram 3

UK energy consumption 1950–1990: heat supplied (billion therms)

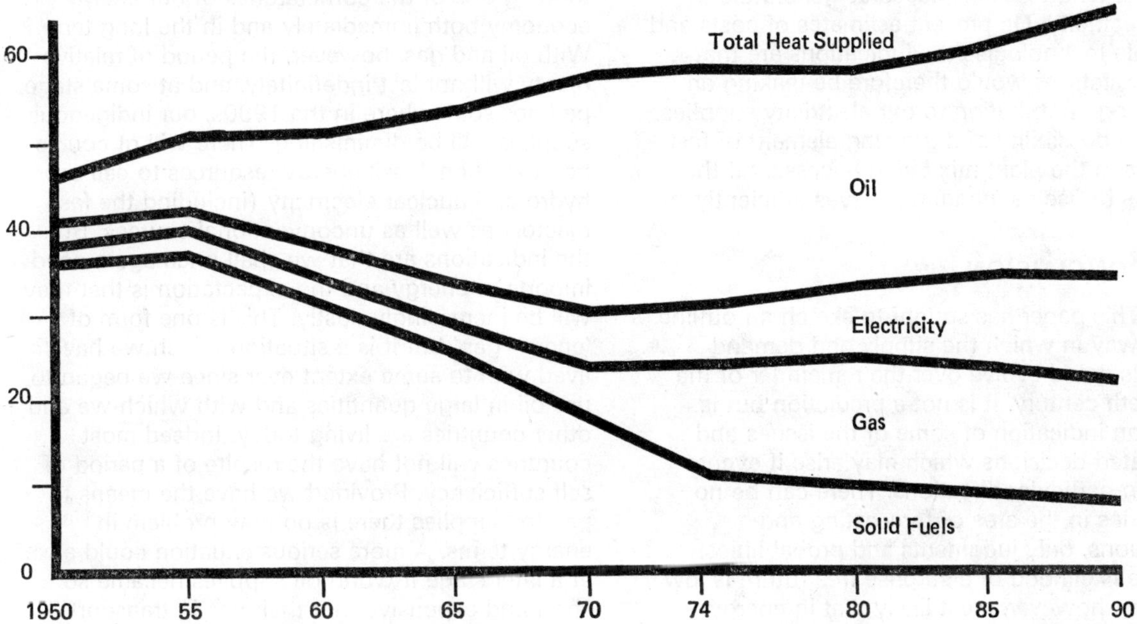

to expand in Europe, North America and the developing countries, then these forces will certainly affect the evolution of the UK economy. The international energy background is discussed in an accompanying paper (page 4).

30 There are many possible and plausible scenarios which can be painted of the UK and world future: with and without growth, with more or less energy, with and without discontinuities, and so on, but this paper looks only at where current trends and policies may be leading. Moreover the very long lead times involved in energy planning mean that decisions taken today (to sink a pit or build a power station) will still be affecting events well beyond the turn of the century.

31 It has already been noted that by the year 2000 energy consumption could be around 550 mtce, with possibly up to a further 70 mtce of hydrocarbon fuels being required for non-energy and petrochemical purposes. Oil from the North Sea, on the basis of present knowledge of reserves, would be approaching exhaustion, but demand for oil would be continuing to rise—much of it in the transport sector unless major technological changes towards electric vehicles had been achieved. On the same assumption natural gas from the North Sea would also be in decline, limited in its use to premium markets and supplemented by more costly synthetic gas made from oil or coal. On the other hand there is a strong possibility that further quantities of oil and natural gas will be discovered, and further investigation is needed to establish the range of coal production in the later years of this century. The demand for coal in traditional uses especially for steam raising would probably be declining although a substantial market would remain in electric generation. Coal's future in the year 2000 and beyond would lie in the development of new conversion processes.

32 In the field of electricity use and generation there are great uncertainties. Growing historically at twice the rate of energy consumption as a whole its share has steadily increased and could, if this continued, represent well over half of total primary fuel demand by 2000. The proportion of heat supplied would be much lower but it poses the question of how this electricity is likely to be produced, how it is likely to penetrate the different

markets, and what structural and environmental issues would be involved. Unconventional sources, such as the wind and waves, could be making a modest contribution to electricity supplies. At the same time the value of coal and oil in other uses will be increasingly important, and they will tend to be displaced as generators of bulk electricity. On present estimates of costs and available technologies the indications are that nuclear stations would therefore be making an increasing contribution to our electricity supplies, with the possibility of a growing element of fast reactors in the plant mix being necessary if the world is to use its uranium reserves efficiently.

VII CONCLUSION

33 This paper has sought to sketch an outline of the way in which the supply and demand for fuels could evolve over the remainder of the twentieth century. It is not a prediction but is rather an indication of some of the issues and associated decisions which may arise if events move in particular directions. There can be no certainties in the area of forecasting and projections, only judgments and probabilities, and the likelihood of a surprise free future is low. It seems, however, most likely that in energy terms the UK will be enjoying a period of relative abundance through the 1980s and perhaps beyond. We shall have more energy available from our own resources than will be needed for our own use and our problems will be how to use it and in what proportions, how much to conserve for the future and how much to export. But although it will be plentiful it will not be cheap for the great bulk of our needs.

34 We are fortunate that our ample reserves of coal, sufficient to last for some 300 years at current rates of consumption, are capable of forming one of the cornerstones of our energy economy both immediately and in the long term. With oil and gas, however, the period of relative plenty will not last indefinitely, and at some stage, perhaps somewhere in the 1990s, our indigenous supplies will be diminishing. There will of course be other, non-fossil energy resources to call on: hydro and nuclear electricity (including the fast reactor) as well as unconventional sources. But the indications are that we shall once again need imports of energy and the expectation is that they will be increasingly costly. This is one form of 'energy gap' but it is a situation which we have lived with to some extent ever since we began to use oil in large quantities and with which we and other countries are living today. Indeed most countries will not have the respite of a period of self sufficiency. Provided we have the means to pay for supplies there is no new problem in energy terms. A more serious situation could arise at a later stage if world oil supplies became so short and expensive that their use in transport became impossible and if by that time cheaper and more plentiful supplies of other fuels and new technologies had not enabled the economy to adjust. The developments in this direction are discussed in the paper 'Energy R & D in the United Kingdom' (see page 64).

Energy prospects (Energy Research Group, Cambridge University)

This paper is published separately as Energy Paper Number 12 by Her Majesty's Stationery Office.

Coal

Long term prospects for coal (Joint Policy Advisory Committee of the Coal Industry)

ENERGY PROSPECTS AND POLICY

1 It is important that decisions in the energy field are taken against the background of long-term energy prospects. Although the uncertainties are very great there is a strong case for saying that:

(a) The world's scarcest and most valuable energy resources, namely oil and natural gas, are being the most rapidly depleted; so that before the end of the century global reserve limitations will at least restrict growth in supplies to less than the growth in potential demand, which will cause 'real' prices to rise substantially by the year 2000.

(b) By the 1990s North Sea oil and gas could have passed their peak and be declining at a time when renewed imports of primary energy could be very expensive. Even with a significant nuclear programme and continuing conservation, there is likely to be scope for significant expansion of British coal in the long-term if we wish to preserve our self-sufficiency with economic advantage.

2 It seems clear that UK self-sufficiency in primary energy is likely to be achieved by about 1980, continuing for a decade or more and with considerable potential for export. In these circumstances it seems important to decide the right balance batween export and conservation of North Sea oil and gas. Perhaps even more crucial will be the need to avoid a false sense of security. There is a danger that self-sufficiency will create an illusion of continuing energy surplus, which will militate against the long lead-time energy measures which require continuity, namely coal, nuclear power, and energy conservation. For example, it can take up to 10 years to bring a new colliery to full output.

3 In the development of energy policy, UK coal will be of great importance to the end of the century and beyond since:

(a) in the period to 1980 coal will reduce the imported oil burden and will hasten the achievement of energy self-sufficiency;

(b) in the 1980s when self-sufficiency has been attained, a large coal industry will provide continuing benefit to the balance of payments by giving the country the opportunity for energy exports. It should be noted that the balance of payments (and therefore the economy) benefits as much from a ton of oil released for export as a ton of oil import saved. Coal will also provide the country with the option to conserve North Sea oil and gas by controlled depletion;

(c) In the long-term, coal is likely to become increasingly important. It is probable that the 'technically recoverable' reserves of British coal could amount to some 45 billion tons—equivalent to over 300 years at current rates of production and probably some five times greater than the ultimate reserves of North Sea oil and gas.

Coal production

4 The present Plan for Coal is designed to provide over 40 million tons of additional capacity at existing and new mines by the mid-1980s which would enable the industry to expand towards the potential demand of 150 million tons. About half the projects necessary to implement this programme have already been approved and planning permission has been given for the development of the Selby coalfield. However, the exhaustion of the oldest mines will continue beyond the mid-1980s, so that it has to emphasised that Plan for Coal must be seen as part of a continuing process. Coal is an extractive industry seeking net expansion in the long-term, and the appropriate commitment of resources will be required to create new capacity beyond the period covered by the present Plan for Coal.

5 While it is not possible to give precise forecasts of UK coal output in the long-term, since this will depend on the results of continuing exploration effort and the scale of subsequent investment, ultimate coal resources are so large that we are confident that the greatly increased level of coal exploration already achieved will establish economically recoverable reserves

sufficient to enable UK coal output to be increased in the long-term.

6 While we are convinced of the strong economic case for new coal capacity, the burden of interest charges, under the present methods of financing, will cause considerable difficulties during the interim period when the industry is seeking to build up new primary capacity rapidly, following a long period of low investment. This problem is recognised in the *Final Report of the Coal Industry Examination* in which it was stated that the Government would be ready to consider dealing with any special problems of financing that arose.

7 The coal industry recognises that although there is great potential for long-term expansion, the markets available for UK coal will also be conditioned by the industry's own performance. Although productivity in the recent past has been adversely affected by increased development work and the recruitment of additional labour, which takes time to train and deploy on productive work, more fundamentally, we believe that the industry is showing the delayed adverse effects of the low levels of investment in major projects during the 1960s, which has resulted in there being very little efficient new and reconstructed capacity to replace the inevitable deterioration in some of the older pits.

8 However, for the future there is no reason why productivity should not rise substantially since:

(a) there is considerable un-tapped potential from existing technology, as evidenced by the very high levels of performance achieved by the best faces and collieries in the industry. Also we believe that an early improvement might be obtained by the introduction of a more effective incentive scheme (as recommended in the Tripartite Coal Industry Examination report);

(b) the wide difference between the productivity possible at new collieries and the limitations of productivity at old collieries of limited life make it clear that the development of new capacity in replacement of the exhaustion of the oldest pits will make a material difference to the industry's productivity by:

(i) creating mining environments in which high coal face outputs can be achieved,

(ii) designing for high productivity away from the coal face,

(c) although coal-getting operations are now fully mechanised, further technological progress is taking place—the main emphasis being on bringing together into integrated systems the best of present techniques, and on the extension of remote and automatic control. To further this the Board have significantly increased their research and development expenditure on mining.

Coal markets

9 If coal is to play an increasing part in energy supply, it will be essential for the plant and technology to be available for it to be used efficiently. As conventional oil and natural gas become more expensive, the case for increasing the use of coal for steam raising will be strengthened, yet much of the present steam raising plant using coal is old and relatively inefficient, so that a major effort will be required in this area.

10 At present the main steam raising market for coal is power stations. At the joint meeting held with the electricity industry and the unions on the February 20, 1976, the Board pointed out that:

(a) it is now 10 years since work started on the last new coal-fired power station to be ordered, and that in the period to 1980 ten new power stations will be commissioned, none coal-fired, so exacerbating the present position in which there is insufficient modern coal-fired plant to convert coal into electricity at high efficiency, placing coal at a commercial disadvantage compared with other primary fuels;

(b) after 1985, unless new coal-fired capacity is introduced, this position could get worse. We believe that all major fossil stations ordered from now on should be capable of burning coal, and that Drax B and West Burton B should be the first new conventional station ordered and should be started as soon as possible.

11 General industry also represents a major growth potential for coal, particularly in the larger scale installations. Coal could also make considerable impact on industrial generation of electricity and the development of combined heat and power schemes. The Board recognise that high priority will have to be given to the provision of relatively low capital cost coal-burning plant, and the introduction of fluidised bed combustion.

12 The demand for coking coal should continue to be strong for the foreseeable future. The major task for the British coal industry in this field, will be to continue to develop ways of widening the range of coals for the production of high quality blast furnace coke by blending and other means (including formed coke), so as to make the

maximum use of indigenous coal and to keep to a minimum the imports required on quality grounds.

13 In the domestic market, notwithstanding the substantial further expansion of gas, the Board see a continuing market for selected solid fuels. We shall continue to give attention to the development of new appliances, improvements in distribution, and will be examining the increased use of coal in district heating.

14 In the longer term, (say from the 1990s onwards), the prices of crude oil and natural gas could well justify the gasification and liquefaction of coal. It will be important for the UK to have access to this technology, particularly using integrated processes which will reduce total capital costs, greatly reduce waste heat and improve the overall energy utilisation involved in current large-scale power generation. The National Coal Board's (NCB) views on the commitment to Research and Development resources on coal is set out in the Annex below.

Imports and exports

15 It has been suggested that imported coal might provide a significant contribution to UK energy requirements of the long-term. Our view is that a strategy based on coal imports would be mistaken, except insofar as imports might complement indigenous production, since:

(a) for the foreseeable future the price of internationally traded coal will be conditioned by the world oil price, and therefore will show little advantage over expensive oil imports in the long-term;

(b) supplies of imports on a continuing basis for the long-term are likely to be uncertain in view of the other demands which will be placed on coal worldwide;

(c) large scale coal imports would represent a significant burden on the balance of payments.

By the same token, these circumstances could well increase the opportunity for export of UK coal to the Community, where the import requirement seems likely to increase from the present level of some 40 million tons to some 60 millions by 1985 and still further thereafter.

CONCLUSION

16 Coal has by far the largest ultimate reserves of fossil fuels in the UK, and will assume increasing importance in the longer term. For the potential to be realised, there will need to be not only efforts by all within the industry to achieve substantial increases in productivity, but also an appropriate and continuing commitment of resources both for production and utilisation of coal.

17 This memorandum is submitted by the coal industry's Joint Policy Advisory Committee, representing the National Coal Board, National Union of Mineworkers, National Association of Colliery Overmen, Deputies and Shotfirers, and the British Association of Colliery Management.

ANNEX: RESEARCH AND DEVELOPMENT OF COAL UTILISATION TECHNOLOGY

The National Coal Board's existing research programme at their Coal Research Establishment, while adequate to preserve familiarity with the technology, is modest in scale (amounting to some £5 million per annum of expenditure). What is required is a larger programme, including the provision of pilot plants as the first stage towards scaling up the technology to commercial use. Because of the scale of expenditure involved and the wider issues of energy policy raised, the Board consider that a major portion of the funding for such a programme should come from Government and/or other organisations and agencies whose activities would be affected.

In January 1974 the Board drew up for submission to Government proposals for a national programme of development on coal conversion processes. This was considered by the Tripartite Coal Examination during 1974. The main outlines of the proposed programme estimated to cost some £40 million at December 1973 prices were:

(i) a pilot plant demonstrating fluidised bed combustion for electricity generation on a 20 MW scale. The Board consider that this method of combustion has great advantages and could prove to be competitive for smaller mid-merit stations where the economics of nuclear power are likely to be unsatisfactory. Fluidised bed combustion will also have considerable potential for industrial steam-raising;

(ii) coal liquefaction by solvent extraction and hydrogenation of coal;

(iii) a gasification plant designed to produce synthesis gas from coal as the basis for hydrogen manufacture, substitute natural gas (SNG) or methanol;

(iv) gasification to produce a low Btu gas for consumption in advanced power generation cycles;

(v) pyrolysis plant designed simultaneously to produce gas and oil from coal in varying proportions;

(iv) further development on conventional and new fuels for metallurgical work.

In parallel with this programme of plant and unit process development, the programme proposed by the Board also made provision for analytical studies on how the separate processes could be integrated in various ways in coal refineries designed to make maximum overall use of the fuel resources available. Such an approach would reduce total capital costs, greatly reduce waste heat, and in particular would represent a very great improvement in overall energy utilisation from that involved in current large-scale power generation.

The *Final Report of the Coal Industry Examination* recommend that the Government should be prepared to make a contribution to the financing of three projects, namely:

(a) fluidised bed combustion;

(b) coal liquefaction by solvent extraction;

(c) pyrolysis.

Agreement has been obtained for Government involvement with a fluidised bed combustion Demonstration Unit at Grimethorpe under the auspices of the International Energy Agency and under the direction of the NCB. Government assistance for other projects is still being pursued.

Gas

United Kingdom energy prospects and the plans for gas (British Gas Corporation)

INTRODUCTION AND SUMMARY

1 The discovery and exploration of natural gas in the North Sea have revolutionised the role of gas in Britain's energy economy. Gas sales have increased fourfold, primarily at the expense of imported oil, saving £1,000 million a year on our import bill. Gas now contributes the largest share of energy supplies to the domestic market (40 per cent in 1975), and over one-fifth of total final energy consumption (nearly 30 per cent if transport is excluded). Further supplies of natural gas will be coming ashore from next year.

2 During 1975, however, total energy consumption in the UK has fallen substantially. This fall reflects the sharp increase in oil prices, the Government measures for fuel saving, and the deep economic recession. The fall has led to lower oil imports, but also to growing stocks of coal and to surplus electricity generating capacity. The coal and electricity industries have been advocating an artificial stimulation of their markets at the expense of gas. Their case rests on the presumption that they will be required to meet an energy shortage in the longer term, and makes the incorrect assumption that gas is following a policy of rapid depletion.

3 British Gas is vitally concerned that the right decisions should be taken on future energy policy. The nation should reap the full benefit from the enormous capital expenditure now being undertaken to receive and distribute gas from the giant British and Norwegian Frigg gas field, which will increase supplies by a further 40 per cent, as well as associated gas from the North Sea oil fields. Britain's energy resources need to be used efficiently and effectively, minimising waste from conversion losses, inefficient appliances and inadequate insulation. Above all, our fuel industries must make the maximum contribution to the recovery and future strength of the British economy.

4 In this paper we examine Britain's energy prospects for the next 25 years and discuss the economics of exploiting North Sea gas. Given the probable extent of our oil and natural gas reserves, and the effect of higher energy prices and of fuel saving policies on demand, an 'energy gap' in the 1990s does not appear at all likely. There is therefore no justification for taking decisions on gas depletion that would be highly uneconomic and would place an unnecessary and heavy burden on gas consumers. After many years as a net importer of energy, Britain will have a surplus of energy for some time to come. The export of this surplus will bring badly needed relief to our balance of payments and to the economy generally. Exports could well include coal as well as oil if the coal industry can produce at an economic price. Coal production capacity would thereby be retained for the longer term. The depletion policy for gas (and for oil) should give due weight not only to the value of our reserves, but to the economics of production and distribution.

5 The paper concludes by describing the plans of British Gas for the next five years, together with the benefits that these will bring to the British economy, to the 14 000 000 gas customers, and to the 100 000 employees of British Gas. Any distortion of its short-term plans could only be justified if the longer term could be clearly forecast and if it could be shown that the gas industry's present policy conflicts with the national interest. In fact there is no such clear projection to the 1990s or beyond. Nor is there any doubt that the gas industry's current operations are in the national interest in aiding the recovery of the British economy, in further improving the balance of payments, and in reducing the rate of inflation.

An energy gap

6 Discussions of energy supplies over the next 25 years are tending to take as axiomatic an energy shortage in the UK in the 1990s. Statements in the discussion at the Church House meeting on February 20, 1976, by several speakers appeared to accept, almost without question, an 'energy gap' at that period.

7 In the view of British Gas, such an assumption is very much open to question; certainly we would all be most unwise to allow energy policy decisions which have to be taken over the next few years to be determined by such an uncertain premise.

8 In considering the energy gap premise, it is worth reminding ourselves how much in error long range energy forecasts have always been—a point that was rightly emphasised by Mr Chapple of the Electrical, Electronic, Telecommunications and Plumbing Unions (EETPU) at the earlier Church House meeting. The 1967 White Paper*, for example now nine years old, underestimated 1970 primary energy demand by 6 per cent and overestimated 1975 demand by 9 per cent. Within this total, although gas projections have been largely justified, those for some other fuels were largely wrong (particularly nuclear energy).

9 It is not entirely clear what is meant by an 'energy gap', nor whether all who use the term attach the same meaning to it. Britain has had for many years a 'gap' between its energy requirements and its indigenous production which has been met by imports of oil. This gap has been reduced in recent years by the discovery of natural gas, which has led to a big saving on our oil import bill. By the early 1980s North Sea oil will eliminate the gap, and even make Britain a net exporter for a number of years—with great benefits to our balance of payments. When North Sea oil and gas supplies decline (which, as we argue below, may well not be until the next century), we shall have a choice of sources for replacing them. The correct course must depend on the relative costs of the various energy sources at that time. Since present projections could be wildly wrong, we should avoid firm commitment to a particular course until lead times make it necessary, while concentrating on research and development work, to make a wide choice of alternatives technologically realistic.

10 We expect energy demand trends to adjust to the change in energy prices and the government's fuel saving policies. There are great pressures, and great opportunities, for using energy more efficiently and economically and adapting our industrial structure to less energy-intensive techniques. There are already signs that adjustment to the new energy situation is taking place. We also believe that the era of significant discoveries hydrocarbons around the shores of Britain has not come to an end. Only if all these expectations prove wrong is a shortfall in indigenous energy likely to appear in the 1990s.

Fuel Policy. HMSO 1967. Cmnd. 3438

Energy demand trends

11 During the 1960s, UK final energy consumption grew on average at $1\frac{1}{2}$ per cent a year. Over the same period, the real price of oil fell by about one-third. Throughout the world, the low price of energy was reflected particularly in the expansion of energy-intensive industries (metals, chemicals) and of energy-intensive services (road and air transport). In our view, the fourfold increase in oil prices, together with the increasing public awareness of the need for efficient and economic use of energy and of the opportunities for energy saving, will substantially alter these trends. In addition, as the recent Open University *'Critique of the Electricity Industry'* has pointed out, account must be taken of the approaching saturation of some markets, particularly the domestic market. As a result we expect a much lower rate of growth in final energy consumption over the next quarter century, even with a rate of economic growth similar to, or rather higher than, the historical rate.

12 Energy consumption in 1975 was at a low level because of the economic depression and 1974 therefore provides a better base for future projections. In 1974 primary energy consumption was 331 million tons of coal equivalent (mtce). Final energy consumption—allowing for conversion losses—totalled 229 mtce (58·5 thousand million therms). If growth in energy consumption were to continue at the same rate as in the 1960s, our primary energy consumption could reach 585 mtce by the end of the century. However, our estimate—taking account of the higher price of energy, the substantial opportunities which exist for energy saving, and the saturation of some markets—is that final energy consumption will not require more than 500 mtce of primary energy, and quite likely only 450 mtce. In other words adjustment to the high cost of energy is likely to reduce demand for primary energy by 85 to 135 mtce by the end of the century. Any shortfall in indigenous supplies will therefore be significally less than is suggested by the projection of past trends.

Energy supplies

13 Britain is now a four-fuel economy. There are possibilities in the long run of additional sources of energy—particularly renewable resources—but we do not think they should be relied upon to make a significant contribution in this century. Our coal reserves are ample—sufficient for 300 years was a recent Ministerial estimate. Oil and gas reserves, although quanitiative estimates are subject to some uncertainty, are both undoubtedly very considerable

14 The latest *Brown Book (Development of the oil and gas resources of the United Kingdom 1976)*, prepared by the Department of Energy, published by HMSO, states that reserves of oil in existing licensed areas could total between 2,300 and 3,200 million tonnes', and 'taking into account estimates of reserves of oil in areas of our shelf which are already designated but not yet licensed the total reserves could reach 4,500 million tonnes' (fifty times our current consumption). These reserves 'may well . . . be increased once the extent and nature of these further UK areas [ie areas where the boundaries with other countries have not yet been determined] are known'. The figures quoted include proven oil reserves of 1,350 million tonnes—nearly 30 per cent more than in last year's *Brown Book*.

15 Oil production rates, which the *Brown Book* gives for the 1980s, are equivalent to 170 to 225 million tons of coal (mtce), which will permit a valuable export trade. No figures are quoted in the *Brown Book* for the 1990s. Nevertheless, it is not unrealistic to suggest that, allowing for further discoveries, production could remain above 170 mtce, so that Britain could well remain self-sufficient, or even a net exporter.

16 For gas, the *Brown Book* shows proven reserves at 28·7 trillion cubic feet (tcf), probable at 11·5 tcf and possible at 10·3 tcf—a total of 50·5 tcf. Of this total, over half is already under contract to British Gas. The figures exclude the natural gas already used since 1967. They also exclude the part of the giant Frigg gas field which is in Norwegian waters but contracted to British Gas. Even though natural gas accounted for 97 per cent of total gas supplies, and these supplies were four times larger than ten years ago, the proven reserves of natural gas rose, last year, by more than the quantity consumed.

17 Estimates of further discoveries of gas which would add to the reserves' figures quoted are particularly difficult to make, largely because of the varying ratios of gas to oil in 'associated gas' fields. However, as with oil, there are large areas around the coasts of the UK as yet unexplored. There must be a presumption that significant finds are still to be made in these areas, as well as in those areas which are the subject of continuing exploration activity at present.

18 Gas at present contracted to British Gas—particularly the large Frigg gas field and associated gas from the Brent oil field, will raise availability of natural gas in the early 1980s to about 90 mtce (23,000 million therms) compared with about 55 mtce a year at present. Gas will then be supplying about one-third of total energy requirements on a heat supplied basis, or about a quarter of primary energy requirements.

19 The reserves of natural gas described in paragraph 16, if extracted after 1980 at a constant rate, would last into the next century. In practice, a constant rate of extraction throughout the life of a gas field is not possible. For each field there is a pattern of supply consisting of a build-up period, followed by a 'plateau' for some years and then a slow decline (a 'long tail'), since it is not economic to install more and more compression capacity as gas is extracted and consequently the pressure in the field falls. However, it is our reasonable expectation that additional discoveries will be made which will at least offset the decline from existing fields and maintain a high level of supplies to the end of the century.

20 The policy of British Gas on the depletion of our reserves is to conserve gas by giving continuing priority to the 'premium markets' (ie domestic usage, other space heating and industrial processes where a high grade fuel is required). However, account has also to be taken of the need to operate economically and to maintain operational flexibility. Economic factors include the need to make the best use of scarce capital and manpower resources by achieving a good load factor overall (the load factor of domestic supplies is only about 35 per cent). The importance of use of resources is not confined to British Gas—the North Sea producers likewise are subject to economic pressures to utilise their pipelines at high load factors. Operational flexibility is required to meet changes in loads resulting from the great variation in temperatures in the UK and to provide an acceptable level of safety and security of supply.

21 For these reasons, it is the policy of British Gas to sell a limited quantity of gas to non-premium customers on an interruptible basis. The alternatives of very large storage capacity, even if practicable, and under-utilisation of transmission capacity, would be very costly indeed. But, we repeat, the primary objective is to meet premium demand and to restrict non-premium sales to the level required for operational reasons.

22 Given this policy on natural gas depletion, and with further discoveries of gas, there is every expectation that natural gas supplies will hold up past the end of the century. Whenever supplies do decline, it will not be sudden, and it will certainly be technically possible, and should be economic, to continue premium gas supplies by supplementing natural gas with substitute natural gas made from

oil or coal. British Gas is the world leader in the technology of substitute natural gas (SNG), and is currently selling its 'know-how' to many countries, while also continuing its own research and development.

The total UK picture

23 By the year 1990, based on current reserve estimates, natural gas should still be supplying about a third of final energy consumption, or a quarter of primary energy. Indigenous oil production is likely to be at least adequate to maintain current levels of consumption, leaving the remainder to be supplied by coal, nuclear and other primary electricity. Beyond 1990 there is the possibility of a gradual decline in indigenous supplies of oil and gas, although this is unlikely to occur until after the year 2000. The extent to which, if a decline does occur, it should be covered by imported oil, by additional coal consumption, by more nuclear capacity or in other ways, is very unclear. There is a strong case for increasing the options open to us, for example by expanding research and development into all aspects of energy production and utilisation. In addition, if coal can be produced at a world economic price, then capacity can be maintained in case of future needs, by exporting the surplus over current requirements. It would be most unwise to enter into commitments which can be avoided for the present, in view of the great uncertainties of the future. This caution must apply to proposals to expand coal capacity beyond the level at which costs can be covered by a competitive price, to the adoption of highly uneconomic depletion policies for gas or oil, or to investment on a dramatic scale in nuclear power.

24 Although Britain appears likely to be in a much stronger energy situation than any other major western country by this time, it is impossible to isolate its energy situation from that of the world. OPEC may well continue to exert a major influence on the world energy scene for many years yet, but its very success in transforming world markets from low to high price is likely to have a profound effect on demand and further exploitation of energy sources. Most estimates of world fossil fuel reserves were prepared before the recent huge rise in oil prices. Nevertheless, proven crude oil reserves for the world are equivalent to about 30 years' current production. The natural gas reserves/production ratio is even greater. In world terms offshore exploration for oil and gas is still in its infancy, and with the stimulus of higher prices there must be strong hopes of major discoveries available for supply into the next century.

The attack on gas

25 At the previous Church House meeting, the National Coal Board and the Central Electricity Generating Board were united on one issue. Gas was too cheap and was appropriating too big a share of the market. This argument, even were its conclusion politically, socially and economically acceptable, was based on a series of contradictions and false premises. First, it was alleged that there would be an 'energy gap' in the 1990s and therefore gas must be conserved. We have shown above that, while a shortfall in indigenous energy is possible, it is unlikely. It is more probable that there will be further discoveries of hydrocarbons, and even a surplus of energy for export is a possibility. Secondly, it was implied that British Gas is following a policy of rapid depletion. This is not so. Short-term financial expediency would suggest a much faster rate or depletion—involving, for example, massive sales of gas to power stations comparable to those in USA and most European countries. But British Gas has followed a policy of preserving natural gas as far as is practicable for the premium markets. Thirdly, it is implied that a policy of greater conservation would lead to a smaller loss of electricity and coal markets. This is not the case. Slower depletion would principally mean lower gas sales in the bulk energy market (although this would present economic and operational problems, as explained earlier). In these markets gas would be replaced by oil. Electricity and coal (directly and indirectly) are losing out to gas in the premium markets, and would continue to do so if depletion were slower. They are losing because of the much greater overall efficiency of gas for premium uses. This gain for gas is a gain for the nation, and is rightly reflected in lower prices for gas to the consumer. Gas is the only fuel which can offer improved heating for the lower income family without increasing the demand for primary energy.

26 The Price Code has imposed an artificial restriction on the price of tariff gas. Gas sold under contract has, however, subject to Price Code rules, been kept at market related prices. Modest increases in gas tariffs, below the expected rate of inflation, would enable British Gas to build up its financial reserves (as required by the Gas Act) and to finance a greater part of investment from its own resources, thereby reducing the public sector borrowing requirement with its inflationary effects. These moderate increases, correctly reflecting longer term cost movements, would still leave gas an attractive proposition for the domestic customer —particularly if the prices of other fuels correctly

reflect their future cost, including the replacement of capital. The opportunity to cover basic fuel requirements for cooking and space heating at a price which they can be expected to afford, should not be denied to consumers by artificial increases such as a tax on gas.

Planning for gas

27 British Gas is planning in the framework of the energy market prospects described earlier. For the next few years planning is dominated by the expected arrival of gas from the Frigg field next year—which will increase supplies by 40 per cent—and associated gas from the Brent oilfield two or three years later. This increase in supplies is requiring large capital expenditure (some £1,500 million over the next five years), the greater part of which is already committed. Planning is as robust as possible to variations in future supply prospects—and the extensions to the transmission system are designed—in the confident expectation that further supplies will maintain deliveries from the northern part of the North Sea for many years to come.

28 Future plans are strongly influenced by the conditions of the supply contracts. The Frigg field requires heavy investment by the producers. It is 220 miles from shore and in 400 feet of water—and part of the field lies in Norwegian waters—factors which have been decisive in determining the build-up, the rate of depletion, and the load factor of supply. Similarly, in the case of Brent, the pattern of delivery is determined by the rate of extraction of the oil, rather than the requirements of the gas market. These conditions are also likely to apply to other future supplies, for example, to the gas to be delivered by the recently mooted 'gathering pipeline'; if this gas is not taken as the oil is extracted it will have to be wasted by flaring.

29 There are considerable marketing problems in matching year by year the development of sales to the likely pattern of future supplies, particularly taking into account the uncertainties of arrival dates. British Gas will continue to concentrate on premium sales, but additional interruptible non-premium sales will have to be made in order to absorb supplies as they become available, and to meet the required load factor.

30 The biggest of the premium markets is the domestic market. Over the past twelve years the total amount of heat supplied to the domestic energy markets has remained virtually unchanged. Yet, over this period, domestic sales of gas have risen by 4,300 million therms, or 280 per cent. There has been an unparalleled development in heating standards, not only in new developments but in existing properties. The number of households has grown by 2·7 million, and the proportion with central heating has increased from 11 per cent to 49 per cent. Four million gas central heating installations have been brought into use, together with 7 million main living room gas fires. The savings achieved through replacing other fuels by gas have covered this enormous increase in heating standards. It is clearly highly desirable that these trends should continue; consequently, after allowing for the gradual saturation of some markets, a further increase in domestic gas sales of around 40 per cent is expected during the next five years. The continued policy of some housing developers, particularly in the public sector, of installing electric heating schemes which, whether using imported or indigenous oil or 'coal by wire', utilise primary energy at efficiencies of only 25–30 per cent, is to be deplored. It involves a waste of resources and high running costs for the occupiers of the properties.

31 British Gas has some 14 million customers. A central objective of future planning is therefore to ensure a continuing supply to these customers as far ahead as one can see, under safe and secure conditions. This objective requires not only the proper marketing and utilisation of gas, but an efficient and profitable industry with a continuation into the future of past improvements in productivity. The plans of British Gas are therefore concerned not only with the major strategic issues of supply and demand, but with the detailed management of the industry, and with achieving the close co-operation of employees in improving efficiency and quality of service to the customer.

32 The management of the industry's finances is of key importance in future planning. The heavy programme of capital investment must be justified by the financial return, and it is desirable that a high proportion of new investment should be financed from internal sources without recourse to external borrowing. The future stability and strength of the Corporation must be assured through the development of a satisfactory level of financial reserves as required by the Gas Act 1972. Prices should allow the industry to meet these financial objectives and should reflect present and future costs. If these principles are followed, prices will remain attractive to the consumer and encourage the development of sales which will achieve substantial resource savings for the country.

Electricity

The prospects for electricity (Central Electricity Generating Board)

The role of electricity in the modern world

1 There are many problems in securing the country's future energy supplies, but one thing is certain—electricity will continue to play a vital role. Its unique combination of convenience, flexibility, and versatility make it essential to modern civilised life and to the advanced industrial technology which supports it.

2 The production side of the Electricity Industry is a processor of fuel and uses a wide variety of primary energy sources in order to make its product. Many of these sources, such as uranium or residual oil or certain qualities of coal are not easy to introduce into the economy in large quantity and in an environmentally acceptable way except via the route of electricity generation.

3 Electricity must inescapably reflect the costs and reliability of its primary fuel supplies and, on a heat content basis, is bound to cost more than these supplies. Electricity sells, and always has done, because its virtues allow it to command a premium in the market place. The customer decides whether he thinks that using electricity for some particular purpose is worth the price.

4 Although the Industry's main product is electricity, exhaust heat is available at power stations and the Industry is eager to collaborate with any organisation or industry which is able to make use of this heat.

5 Clearly the role of electricity must be central to any discussion of fuel policy. Some policy issues are relatively short term; other issues arise in planning for the future. The problems which arise in these two areas are quite different and are discussed separately.

Short term problems for electricity

6 Electricity has inevitably suffered from the oil crisis and from its consequential effects. In the three years up to the end of May 1976 the cost of power station heavy fuel oil rose by 280 per cent and in the same period the pithead price of National Coal Board (NCB) coal supplied to the Central Electricity Generating Board (CEGB) rose by 180 per cent. Fuel costs account for more than one half of the CEGB's total costs and there is no way in which the Industry can avoid passing on a large proportion of these increases in fuel costs to the ultimate consumer via higher electricity bills. The price increase has a fairly rapid effect on demand —something reinforced by the major economic recession and the increased awareness for energy conservation. The demand for electricity is now lower than it was three years ago.

7 The Generating Board has faced the problems caused by these world-wide and nation-wide events. It has successfully negotiated with its Unions early closure of surplus capacity in 47 power stations, affecting some 5000 employees, many of whom will be found jobs within the Industry through redeployment and retraining.

8 The other fuel industries have been similarly affected and, indeed, the situation in one industry reacts on that in another.

9 Both North Sea gas and oil are national assets which have greatly increased in market value since the oil crisis. In the case of oil, Government policy is to redistribute part of this value by means of the petroleum revenue tax and to use it for the benefit of the community as a whole. There is no similar tax on North Sea gas from the Southern Basin.

10 The nation is fortunate in having cheap North Sea gas but its exploitation has inevitably further depressed electricity sales. As a result, the production side of the Electricity Industry has required from the NCB less coal for electricity production than it would otherwise have needed. The Coal Industry is not able to adapt rapidly to this new situation and has continued to produce coal at a rate more appropriate to earlier and more prosperous times. Stocks of coal on the ground are in consequence increasing.

11 The CEGB has, therefore, naturally been under pressure to take measures to burn more coal. It has of its own accord taken action where it is not against the interest of the electricity consumer. For example, it expects to burn in its main oil fired stations in 1976/77 only half the amount of residual fuel oil that it did in 1974/75, and the lower oil burn means a higher coal burn. Additionally it has played its part in accepting higher coal stocks at power stations. Coal industry spokesmen have also argued that the CEGB system needs more high efficiency coal burning plant in order to increase coal burn. To that end it has been proposed that some of the high efficiency oil fired power stations should be converted to coal firing, or at least to dual firing. This, however, misses two points. Firstly, it is not insufficient high efficiency coal burning capacity but the present low level of electricity demand which has reduced the CEGB's requirement for coal; the CEGB already has more than sufficient modern high efficiency coal fired plant to burn the coal should the demand recover. Secondly, the proposed conversions would actually make things worse since the high cost of the conversions and consequent increase in the price of electricity would lead to a further downturn in electricity demand and hence would reduce coal requirements still further.

12 Consequently the CEGB feels that such proposals would not in fact solve the Coal Industry's short-term production problem. Indeed, it does not feel that high coal stocks in themselves are a major problem since they are a valuable national asset which helps to guard against future uncertainties. However, the Board is quite prepared to consider converting some oil-fired plant so as to be able to burn oil/gas. It would then be possible to take regulatory gas supplies at the expense of imported fuel oil and not at the expense of coal.

13 The CEGB takes pride in its achievements in keeping down to a minimum the impact of the substantial increases in the cost of fossil fuels and in inflation generally. It recognises moreover that there are severe social problems associated with the Coal Mining Industry. It is of the opinion, however, that if steps have to be taken to assist the Coal Industry, the costs involved should be regarded as social ones to be borne by the general tax payer and not by the electricity consumer as such. Merely to load additional costs on the latter would be counter-productive so far as the Coal Industry is concerned.

The future prospects for electricity in the UK

14 These shorter term problems will be overcome in due course and the more fundamental task of energy policy is how best to safeguard future supplies of power and energy at a price that people can afford. Now the problem of forecasting future events is notoriously difficult, and nowhere more so than in the energy field. There is little value, therefore, in setting out a detailed estimate of energy requirements or of electricity production in the more distant future.

15 Apart from completing the four Advanced Gas-Cooled Reactor stations, the CEGB is currently engaged in constructing 7,200 MW of new major fossil fired plant, the 1,675 MW Dinorwic Pumped Storage station and 1,200 MW of new gas turbine plant. It is also preparing to build Sizewell B Steam Generating Heavy Water Reactor Station of about 2,500 MW capacity as part of the Government's 4,000 MW programme of SGHWR nuclear plant. These new projects were all planned at a time when future demands for electricity were expected to be higher than they are now anticipated to be. On present forward demand estimates, the CEGB does not need to order any further generating plant until about 1979 or 1980. It therefore has a breathing space in which to consider its future strategy.

16 Turning to the wider scene, the UK is, of course, well endowed with reserves of coal, gas and oil and, by about 1980 when North Sea oil is in full production, can be expected to be in energy surplus for some years to come. But it is a characteristic of all oil or gas fields that, after a certain time, their output declines steadily and rapidly.

17 It follows that, although the UK will be in energy surplus for some years to come, it is necessary to give consideration to the securing of the country's energy supplies towards the end of the century. With all the uncertainties of the future, it is clearly impossible to foresee at this stage precisely how this will be done and it would appear prudent for the country to make a careful appraisal of all the options from the point of view of economics, long-term fuel availability, safety and the environment. These options will certainly include, among others, investment in coal, nuclear power, alternative energy sources and energy conservation. Undoubtedly, electricity will continue to play its part in providing the nation's energy supply, but to what extent depends on just how these various options are developed.

Energy and electricity in Scotland (South of Scotland Electricity Board)

Introduction

Changes in the sources from which energy supplies can be obtained necessarily take place slowly. New sources cannot in general be exploited quickly on a large scale, so major changes cannot be improvised hastily. Planning and preparations for such changes have to be made a long time ahead —often a decade or more.

The development of supplies of oil from the North Sea has required major efforts over many years, and several more years will be required to build up to full production. The development of nuclear sources of energy is taking even longer, and a similar long timescale is likely to apply to the development on any commercial scale of the renewable sources of energy at present under consideration.

When the price of oil rose suddenly and sharply in recent years no ready expedient was convenient to hand. The higher prices had to be paid with the resulting dramatic effects on cost of living, economic activity and balance of payments. It is purely fortuitous that there lies a store of energy under the North Sea sufficient to restore energy self-sufficiency for at least a few years. However, North Sea oil and gas are not permanent solutions. These sources are of limited extent but our energy needs will continue into the future. A new phase of the problems of supplying enough energy to meet our needs will open in the not too distant future and when that occurs we will need sources of energy much larger and more permanent than our most optimistic view of the oil and gas reserves of the Continental shelf. To see what will be required and to have time to make the preparations required to provide for our needs, we must look well ahead and we must ensure that our actions meanwhile are such as to make it possible to achieve the desired ends.

The energy gap

Looking to the future involves many uncertainties and we need to consider a range of possibilities. The top two curves in Fig 1 (p. 37) give two views of the fuel requirements of the UK for energy and non-energy uses, the latter mainly as feedstocks. These two views are based on different patterns of economic activity and other factors which together result in different rates of growth of energy requirements. Both include allowances for restraint in growth due to conservation policies and the high cost of fuels. The rates of growth vary over the period, reducing generally in later years, and broadly assume that the UK standard of living will not be allowed to fall substantially in real terms.

The lower two curves show two estimates of indigenous energy resources, mainly fossil fuels, coal, North Sea oil and natural gas. The actual supplies at the beginning of the period are shown together with optimistic and pessimistic forecasts. The largest influence in the difference between these two forecasts is in the rate of production of oil from the North Sea. This is the largest single source of energy in the period and probably contains the greatest range of uncertainty.

Between these two sets of curves there lie two gaps, one large at present but closing as we go forward to about 1980, the other opening from the 1980s. The first gap is at present being filled mainly by imported oil with a small amount of nuclear energy. The second gap has quite different characteristics and it is potentially much larger. It widens as we go forward and its widening shows no limit. It is literally open-ended. The gap could be closed temporarily by using indigenous oil and gas at higher rates but because these are limited in extent it follows that the gap would grow more rapidly in later years.

The four curves are therefore illustrative. They do not necessarily embrace all possibilities. We are dealing with the difference between two forecasts each of which is hedged with uncertainty. The exact extent of the second gap in our energy supplies in any year is uncertain. But the gap itself is inevitable as is its general shape; all that is in doubt is the scale of time on which it will develop and grow. It may arrive a little later than shown in the diagram and its rate of growth be a little less. It may be with us earlier and grow more rapidly. We need to be in a position to deal with it whatever the timing of its arrival and its rate of growth.

Filling the gap

Much effort has been and doubtless will be expended in seeking to develop renewable resources of energy to use in the future energy gap. The only renewable source employed at present, hydro-electricity, has already been exploited in the UK to about the practical limit and any further development could not contribute significantly.

Tidal power is possible but has limited application and even if such schemes can be made economic they are unlikely to contribute more than a few per cent of the total energy demand.

There is an increasing amount of work being done on the harnessing of wavepower but there are still major engineering problems to be solved and this method would almost certainly be very expensive.

Geothermal sources have little application in the UK and cannot be counted upon to contribute significantly to solving this problem.

Solar energy is most readily exploitable in those areas of the world with least energy demand – that is, the areas with most sunshine. Its use in this country is likely to be restricted to applications for space heating and hot water supplies. Capital cost is at present high and seems likely to remain so and as a source of energy the process would be unreliable and likely to be least useful at the time of greatest need.

There are one or two other small sources, such as the use of methane produced from sewage and burning domestic refuse, and wind power. Their contribution to the energy demand in this country is likely to be a fraction of 1 per cent.

Another approach to filling the gap is to examine the scope for improving the efficiency with which existing energy sources are used. All fuel use involves losses. With most fossil fuels the loss is greatest at the point of use and tends to be high for small-scale use such as in domestic and smaller commercial and industrial applications. With electricity the losses are mainly in the process of production. Some savings can be made and are made by combined electricity and heat supply to factories.

Attempts at more widespread application—eg, District Heating—have encountered problems of high capital cost and high financial risk. The assessment of such projects is complex and cannot be undertaken on the basis of a blithe assumption that the only resource worthy of conservation is fossil fuel. However, the examination in depth of what appear to be the least unpromising projects should continue and in time particular combinations of circumstances which can lead to viable schemes may be identified. The size of contribution that might eventually be made towards bridging the energy gap cannot be foreseen at this stage but it is unlikely to be substantial in the present century.

We are left with the clear conclusion that the principal sources available to make up the shortfall in energy are coal and nuclear power.

We are fortunate in the UK in having large coal reserves, although they are likely to become progressively more difficult to exploit. The need for coal for electricity generation will clearly continue and will be supplemented by demands for coal for gas production and chemical feedstock purposes. For all of these reasons a healthy coal industry will be necessary.

But on the most optimistic assumptions coal cannot fill all of the gap left by diminishing oil and gas reserves. The balance will have to be made up by the only other source which can be seen to have the potential capability—nuclear power. This has been recognised by most other industrialised countries many of whom have plunged into large nuclear programmes.

In the UK we are fortunate in that our fossil reserves permit a more prudent and cautious path in development of our nuclear potential but we have to be constantly aware of the long lead-times necessary for satisfactory development. Our task is to proceed at such a rate in building steam generating heavy water reactors (SGHWR) and commercial fast reactors (CFR) that we can be confident of a sufficient energy contribution at the end of the century, while meantime avoiding over-commitment of financial and engineering resources.

Supplies in Scotland

So far we have looked in this paper at the general energy supply and demand forecasts for the UK as a whole. Although the problems of future energy supplies in Scotland will be generally similar to the remainder of the UK there will be detailed differences which will require to be taken into account. In Scotland coal is more costly to produce than the average for the UK and large areas are remote from supplies of coal and gas and are in the high-price zones for deliveries of oil. On the other hand most of the hydro-electric resources of the UK are in Scotland and the proportion of nuclear output is greater than for England and Wales.

The current energy use in Scotland is shown in Fig 2 (p. 37) with a forecast for the next twenty years. In this case only one forecast is given but the probable extent of each of the constituent sources of energy is shown. The energy sources involved are set out in the order shown so that the hydro-electric, nuclear and appropriate parts of coal and oil may be grouped together to show the sources provided through the medium of electricity. The remainder are supplied to consumers direct. The forecast is based on the following views of supplies from each of the sources.

(a) Gas
Direct supplies of gas to consumers will almost double between 1974 and the mid 1980s and thereafter decline slowly as limitations in supplies of natural gas become clearer and as short-term anxieties to bolster the balance of payments recede and are superseded by long-term concern for preservation of fuels for particular purposes and for supplies of chemical feedstocks. The competitive position of natural gas will be eroded by the need to introduce supplies of more expensive synthetic gas.

(b) Coal
Coal for electricity production may settle at a level of around 8m tonnes a year and there will be a demand in later years for coal supplies as a feedstock for gas and chemicals production. The tendency for prices to increase in real terms will be contained.

(c) Hydro-electric
Constant annual output, varying only with availability of water, is expected over the whole period.

(d) Oil
Direct use is expected to increase slowly to about the end of the century and then to decline with falling production from the Continental shelf and rising world prices.

(e) Nuclear
The balance of energy required will be obtained from mainly nuclear sources. Output is expected to increase over the next few years as the new Advanced Gas-Cooled Reactors at Hunterston B work up to full production but then a period with no growth is expected. This arises from the recent years of low growth in demand for electricity leading to postponement of the need to order new generating plants of all types, including nuclear. Later the growth of nuclear generating capacity will require to become rapid. Continued technical development may be expected to have some off-setting effect against future general inflation of price levels.

Based on these considerations an assessment can be made of the likely requirements for the construction of nuclear plant in Scotland. With the completion of Hunterston B and the Prototype Fast Reactor there will soon be no main nuclear plant under construction in Scotland. But the work load in constructing nuclear stations is expected to start again when orders for the station at Torness are placed, and to increase over the years as orders for further plants follow.

In considering the physical problems of such an uneven programme we should note the present difficult position of UK manufacturers of nuclear plant and electricity generating equipment which arises from absence of orders in recent years. The ability of these firms to supply substantial quantities of plant during the 1980s and 1990s is an essential part of the strategy outlined in this paper and the maintenance of this ability requires early ordering of such plant, albeit in initially modest amounts.

There are further considerations which need to be taken into account. Figure 2 indicates a reasonably steady rate of growth of electricity demand but several factors could intervene to upset this and their consequences could be serious. A recent example has been the very high rate of increase of electricity tariffs brought about by a change of Government policy. Anomalous situations created by allowing prices to get out of step with changing costs can quickly create distortions in the market with results that can last for some time. Another more serious situation will arise if gas sales in the domestic and commercial markets are permitted to grow in the short term to a level much higher than can be sustained economically in the long term. The effect will be to depress electricity growth in the short and medium-term and then to demand a very rapid growth rate later. There will be two damaging long-term effects. The start of a programme of nuclear construction will be further delayed without decreasing the amount of such plant required in the longer term. This will further aggravate the existing difficulties of the nuclear construction industry owing to the low level of current orders. The second effect, to be expected particularly in Scotland, is that the market for coal to power stations will be reduced over the short and medium term. In that event it is to be questioned whether recovery to the level postulated in the forecasts above will be possible. Over-hasty exploitation of natural gas in the near future could restrict the level of production of indigenous coal over a long period. There would surely be overall advantages to be gained if there were a policy for use of gas which took account of long-term consequences.

Energy resource planning is a subject for long-term views and for protecting the long-term prospects from harmful short-term expediencies. Major changes from one source of energy to another, in whole or in part, are slow and require to be planned far in advance. If long-term planning is inadequate in extent or timing, and in inbuilt flexibility, it may be too late to avoid the penalties of our unpreparedness.

Figure 1
UK primary energy: demand and indigenous supply

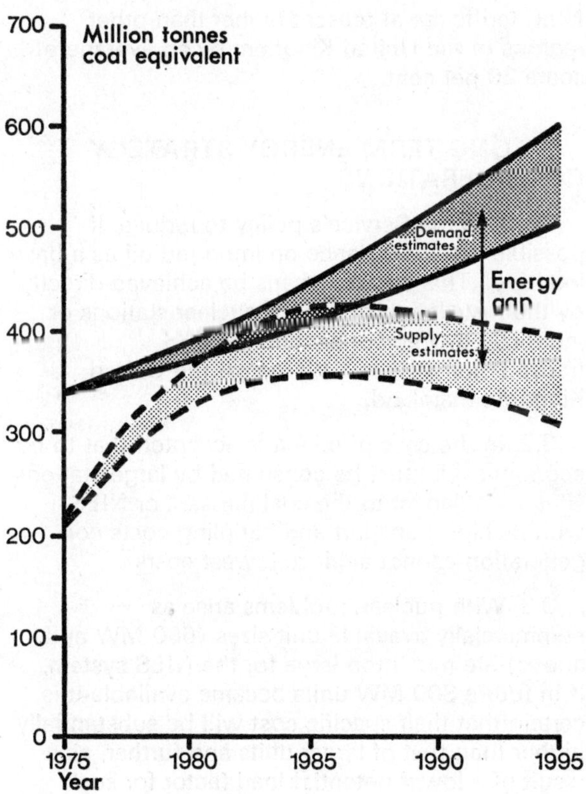

Figure 2
Total energy in Scotland supplied by primary fuels

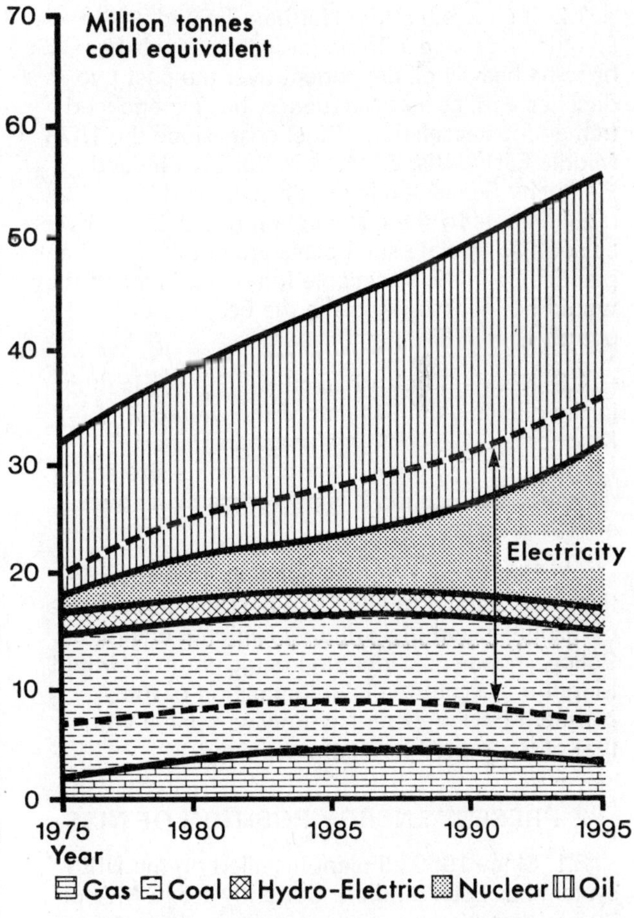

Interconnection – its place in a national energy policy (Northern Ireland Electricity Service)

1 SUMMARY

1.1 The paper states the views of the Northern Ireland Electricity Service (NIES), as an electricity supply authority within the United Kingdom, in regard to the possible formulation of a National Energy Policy.

1.2 It shows that the Northern Ireland Electricity Service with no indigenous fuels has become heavily oil dependent over the past two decades and, as a consequence, has experienced tremendous escalation of fuel costs since the 1973 Middle East Crisis. Whilst it is Northern Ireland Electricity Service policy to diversify fuel requirements by the introduction of Nuclear Power as soon as feasible such plans are not likely to be practicable in the forseeable future, and indeed they would not necessarily make the best use of overall United Kingdom resources.

1.3 Northern Ireland Electricity Service is thus in the dilemma that, as part of the United Kingdom, it is expected to maintain United Kingdom energy prices yet owing to its size and geographical location it is powerless by its own efforts to do so.

1.4 The Paper recommends that, to overcome this dilemma and to allow the best use to be made of resources and fuel within a National Energy Policy, serious consideration be given on a United Kingdom basis to the interconnection of the Northern Ireland Electricity Service system with that of the United Kingdom mainland.

2 PRESENT ENERGY POSITION OF NIES

2.1 Since 1959 all plant installed on the NIES system has been oil fired. Currently 1,545 MW or 76 per cent of generation capacity is oil fired and with completion of a 1,260 MW oil fired station at Kilroot by 1981 the figures will be 2,805 MW or 88 per cent.

2.2 The remaining plant on the system is coal fired and is between 20 and 40 years old.

2.3 To take advantage of economies of scale and to make the best use of a limited number of power station sites, NIES have been forced to employ relatively large unit sizes (at present 200/300 MW). This necessity has led to the adoption of higher plant margins than would normally be the case for a small system.

2.4 As a result of oil dependence and some over-capitalisation because of large plant margins NIES tariffs are at present higher than other regions of the United Kingdom by an average of some 20 per cent.

3 LONG TERM ENERGY STRATEGY CONSIDERATIONS

3.1 It is the Service's policy to reduce, if possible, its dependence on imported oil as a base load fuel. This could perhaps be achieved directly by the installation of coal or nuclear stations or indirectly by interconnection of NIES transmission system with that of the United Kingdom mainland.

3.2 In the case of coal it is accepted that to be economical it must be consumed by large stations situated adjacent to the coal mines. For NIES with its high transport and handling costs coal generation cannot achieve lowest costs.

3.3 With nuclear, problems arise as commercially available unit sizes (660 MW and above) are much too large for the NIES system. If in future 300 MW units became available it is certain that their specific cost will be substantially higher than that of larger units and further, as a result of a lower potential load factor for such plant, it is unlikely that we could achieve the same annual fuel 'burn up' as the United Kingdom mainland. Thus if the capital resources required to establish a nuclear station in Northern Ireland in the medium term future were used to provide nuclear plant in the United Kingdom mainland, not only could additional capacity be provided but increased fossil fuel savings would result both from the increased capacity and from better utilisation.

3.4 Interconnection with United Kingdom mainland using direct current submarine cables is now a practicable consideration and this could achieve economies for NIES and the United Kingdom. The reasons for interconnection between Boards are well known and are basically of the same nature as has led the Central Electricity Generating Board (CEGB), South of Scotland Electricity Board (SSEB) and North of Scotland Hydro-Electric Board (NSHEB) to develop the main transmission and interconnection system on the United Kingdom mainland. The three main advantages in NIES case would be that it would permit.

—a substantial reduction in NIES plant margins with consequent capital savings
—a reduction in overall fuel consumption by reducing NIES day-to-day spinning reserve requirements
—NIES to participate in the longer term in United Kingdom and EEC energy policies by the construction of nuclear stations with power for export.

4 CONCLUSIONS

4.1 Northern Ireland because of its size and geographical location has higher electricity tariffs than the remainder of the United Kingdom and it is unlikely that generation diversification programmes can alter the relative tariff position in the forseeable future.

4.2 NIES tariffs are compared directly with those in the remainder of the United Kingdom and Europe by industrialists who are making investment decisions. It follows that Northern Ireland is at a disadvantage in competing for energy intensive industries.

4.3 Interconnection to the United Kingdom mainland would result in overall United Kingdom capital and fuel savings and ultimately would allow NIES to participate in overall European energy policies.

Flexibility in long-term energy policy (Electricity Council)

INTRODUCTION AND SUMMARY

(i) The Electricity Council is the central body of the 13 federated boards involved in the supply of electricity in England and Wales. It advises the Secretary of State for Energy on matters affecting the supply industry and has a duty to promote an efficient, co-ordinated and economical supply of of electricity.

(ii) The Council acts for electricity boards on certain matters of common interest, such as forecasting and tariffs. It also has specific duties on behalf of the whole industry in respect of capital-raising and expenditure, industrial relations, negotiation of terms and conditions of service for staff throughout the industry, consumer complaints, research and accounting.

(iii) This paper sets out the Council's view on how long-term energy policy should be co-ordinated by the Government so as to enable each energy industry to make its proper contribution towards the country's energy needs.

(iv) Medium term considerations may require high rates of depletion of UK fossil fuel reserves, primarily so as to improve the balance of payments. Nevertheless, the Government ought in the Council's view to seek to reduce its direct intervention in energy decision-taking. To avoid such intervention whilst yet securing a coordinated approach, the Government should set economic and social ground rules as a framework for flexible planning.

(v) The long term aim within those rules should be to achieve adaptability in the producton and consumption of energy at minimum cost, after having due regard for safety and the environment, allowing for a specified degree of freedom of choice of fuel by the user.

(vi) The most beneficial result of this National Energy Conference would be agreement that ground rules which the Government should take the lead on establishing, after due consultation, should be developed as a basis for the energy industries' planning.

Energy industries' planning problems

1 The problems faced by the energy industries in long-term planning stem mainly from the length of time needed to adapt their production capacity to meet changes in demand, when so much depends on the demand's being met. It is as essential for the UK to provide for an adequate future energy supply as for a water supply but the energy problem is international and cannot be solved by the UK in isolation. Just as planning for the UK is complicated by the need to co-operate with other countries, so each energy industry's planning must be linked with, hence complicated by, others in aiming to meet long-term energy demands.

2 Taken together, the oil, gas, coal and electricity industries have tremendous economic importance: 9-10 per cent of national capital expenditure; 5 per cent of the National Income; 3 per cent of the country's workforce. They are major customers for the products of many smaller and numerous large businesses.

3 The importance of the energy industries and the fact that many of them are nationalised, means that they operate within social and economic policies established by Parliament through the relevant legislation. That is why energy conservation, pricing, helping the poor to meet bills, retraining redundant staff, etc, are matters under frequent debate. It is important to settle the social framework for long-term planning just as it is essential to decide the financial framework. Managing such large industries is difficult even when the basic reference points for plotting a course are clear. Too often, Government does not make them clear.

4 It has to be recognised that changing the course of any of the energy industries sets up

stresses that can be very damaging if caused too quickly. A change from expansion to stabilisation of demand or to contraction is something which ought, wherever possible, to be identified well ahead and accommodated by successive adjustments over a long period. This is sometimes as important to those who supply the energy industries as to the consumers and staff of the energy industries themselves. Experience has shown, however, that such radical changes of course are unpredictable by and beyond the control of the industries concerned, being the result of changes in general economic conditions or of political decisions, often taken internationally

5 Any effort to clarify the possible changes which may face energy industries in the long-term, due to economic conditions, national or international political decisions, or other causes beyond the control of the energy industries, is therefore welcome.

Electricity industry's unique position

6 It is particularly welcomed by the Electricity Supply Industry, which has a unique position among the energy industries; nearly every dwelling or business premises in the UK has a mains supply of electricity. That is far from true for gas, nor are coal and oil in that position. Moreover, if ever a consumer is short of gas, coal or oil, he may seek simply to switch to electricity. Under present statutes, the electricity industry is obliged to supply the consumer, which is again not true, for example, for coal or oil. This makes the demand for electricity uniquely dependent upon the supply and demand relationship for each of the other fuels.

7 What is more, even in ten years' time, over 80 per cent of the electricity supplied in the UK will still come from coal, gas or oil. The supply of electricity is therefore directly dependent upon the supply of each other fuel.

8 It is possible to store coal, oil and gas but the storage of electricity is a very limited possibility. This makes the planning of electricity supply capacity dependent on the few hours of highest demand in a period many years ahead, to an extent unique among energy industries.

9 Electricity is expected to be available all the time, in adequate supply, for any use, almost everywhere in the UK, such is the unique dependence of the community upon it. Some people believe such communal dependence means that electricity should be supplied as of right, or at least that not everyone should have to pay the economic price for what they use. Such an issue is for Government to decide but in the absence of a decision, planning is more uncertain.

10 Some seek to take advantage of another unique feature of electricity: people can make it for themselves. Whether or not electricity (or, for that matter, bulk heating) should be generated separately from mains electricity is important to the community, involving questions of amenity, safety, dependability and finance. Such planning decisions should be taken in the public interest on a long-term basis designed to deal responsibly with all four aspects.

11 The electricity industry is unusual in competing so widely with its major suppliers. In 1975/76, 51 per cent of the industry's operating costs went towards the purchase of coal, oil and gas. The coal industry competes strongly with electricity in metal melting, gas does so in domestic heating and cooking, oil does so in non-domestic heating. If such competition were to oust electricity from one of its major markets there would be planning implications not only for the demand for electricity but also for the demand for coal used in the generation of electricity. Thus gas competition with electricity could severely damage the coal industry's best market, especially in the next decade. This competition might reduce the market for coal more than would electricity's direct competition with coal (eg in the home heating market).

12 Oil may be like electricity in this inter-active respect; coal clearly is but gas is only slightly affected in that way. The inter-active effects of the energy industries, when selling, upon the energy industries, when buying, call for sophisticated planning of depletion rates, sound pricing philosophies and strategic marketing policies, if they are to be directed in the long term towards the public interest. The prospect of an abundance of fossil fuels in the short/medium term contrasts with the need to obtain an orderly development of the energy industries to meet expected long-term shortages of fossil fuels; the reconciliation of these two opposing influences may require that consideration should be given to the export of fossil fuels, which would assist the balance of payments, and to their continued availability as an essential chemical feedstock rather than a fuel.

Uncertainties

13 Forecasters know that their estimates will be right only by chance. The fact that no one can

Table 1 Primary energy estimates — UK

Title of estimate *Source*	*Year Est'd*	*Estimated (mtce)* Coal	Gas	Oil	Nuclear & Hydro	*Actual (mtce)* Coal	Gas	Oil	Nuclear & Hydro
NCB *Plan for Coal* 1950	1963 (assumed)	205/ 215	not estimated			194	—	85	4
Ridley Committee, 1952	1961 (assumed)	232	—	14*	2	190	—	49*	3
Fuel Policy White Paper, 1965	1970	175	2	144	17	154	18	146	12

*Excluding oil for road and air transport.

accurately predict the long term energy situation must be taken into account. Some previous primary energy estimates are related to the outturn in Table 1. The market for coal was over-estimated in each instance.

14 In Table 2, errors in estimating demand for electricity are shown to have varied considerably in direction and size during the past 25 years.

Table 2 Percentage by which electricity demand estimates were above (+), or below (—) actual for sixth winter ahead

Estimates made in	%
1950	− 5·8
1955	−10·3
1960	− 3·5
1965*	+35·8
1970	+23·4 (estimate)

*Estimates linked to National Plan

The reason for the errors is not the incompetence of the forecasters but the sheer unpredictability of the determinants of the markets they are analysing. Their job is to reduce uncertainty; they cannot eliminate it. We who are considering the next 25 years are like those who, in 1950, studied the 1975 situation. We are very uncertain, as they were, about technical progress, economic factors, such as the balance of payments, and the security of fuel supplies strategically. The NCB's *Plan for Coal,* 1950, wisely stated: 'The Plan shows what would happen only on certain assumptions about the unknown future. The estimates of future performance are not forecasts, still less targets at which the industry should aim.'

Value of scenarios

15 Just recently, the NCB Chairman warned that a world energy gap could develop within 10 to 15 years. He did so almost simultaneously with the Gas Corporation Chairman's statement that he had seen no evidence at all that there is going to be an energy gap in the 1990s. Amid such uncertainty, the correct timing of major decisions becomes very difficult to achieve. To avoid too early a commitment of resources, investment decisions should not be made until the latest opportunity. Prior to that, however, a wide range of alternatives needs to be examined so that, when the decision has to be taken, it will be mature and unhurried. The alternatives ('scenarios' in current terminology) should be chosen to help us to decide in 1977 what is the range of demands that the energy industries might need to meet over the next two decades.

16 Accordingly, the energy industries should collectively consider how to cope with an upper case, testing the limits to growth of energy requirements up to (say) 1995. As a complement, we should also examine how to cope with a lower case, testing which resources would continue to be or would become unnecessary should energy requirements rise only slowly. At present and for some time to come, we cannot be sure which of a variety of cases will prove the most realistic, so we must make ourselves adaptable enough to meet whichever situation should emerge. That is as vital for the suppliers, as for the staff and consumers, of the energy industries.

Other needs

17 It is also vital for the staff at all levels of these large industries to increase the satisfaction to be gained from their work. To achieve such an important aim requires not only participation in a process of consultation but also, to the degree

appropriate to each task, the power to take decisions and the duty to carry responsibility for them. To achieve this requires an agreed and settled framework which the Government should take the lead in establishing; there should then be less need for Government to involve itself in detailed aspects of the industries' activities; and management would then be able to develop detailed plans more assuredly in consultation with staff.

18 Surprisingly enough, it seems necessary to remind ourselves of the need for integrity and responsibility on the part of the user of energy. If he does not pay for what he uses, someone else will and it is important that such transfers occur only when Government intends that they should. Hence Government needs to state a policy on this matter of fuel debt transfer and fuel subsidy.

Framework for freedom to plan

19 The Electricity Supply Industry considers therefore that, instead of concentrating effort on long-term forecasts which can only be speculative, the Government should aim to produce soon after the consultation in June, a set of ground rules which all energy industries can use as a basis for their development. The rules should be accepted by all interested parties, in common, for a long period so that after 1976 decision-taking may suffer less disruption by short-term pressures. Such pressures often force decisions that are costly in the long-run and gain only a short-lived benefit, in the absence of stated principles. A framework of principles would make for broadly consistent decision-taking in many places at many levels.

20 A cardinal consideration is the amount of freedom that the user of energy is to have in selecting, from time to time, which fuel to use and, as a corollary, the extent to which the energy industries should be able to compete with each other. The issues here are not straightforward; what is the 'best' fuel to use often depends on the individual consumer's judgment of a variety of factors. However, once having made his judgment, or, in the absence of freedom to choose, having had the judgment made for him, he may become 'locked in' to a particular fuel because houses, offices, etc, have a long life. It may be that more research into the design of buildings is necessary, so that they may be more easily convertible from one fuel to another. Because of the problems of long-term prediction referred to earlier, it is not obvious that centralised decisions as to the right fuel to use for any particular purpose will turn out to be correct and it is necessary to weigh this against the extra costs of some duplication of supply capacity if consumers are to have freedom to choose. The permissible degree of freedom of choice is a key feature of energy planning in a democracy.

21 Closely-linked with the energy-user's freedom of choice is the degree of control which the Government may seek to exercise over each energy industry's contribution to total energy requirements. For the medium term, economic considerations, such as the balance of payments, may seem to the Government to drive it to specify rates of depletion of UK fossil fuel reserves. The Government should nonetheless seek to reduce its direct intervention by securing increasing compliance by each energy industry with the broad ground rules, in phases from 1977 onwards. The long-term aim within those rules should be to achieve adaptability in the production and consumption of energy at minimum cost, after having due regard for safety and the environment, allowing for a specified degree of freedom of choice of fuel by the user.

Ground rules required

22 The ground rules should therefore include:

(a) Standards for preserving safety and the environment.

(b) Rates of return required by society from the resources it devotes to energy production, distribution and utilisation.

(c) Principles for pricing and methods of payment for energy consumed.
(NB: They would have to be consistent with (b)).

(d) Degree of freedom of choice of fuel by users, from time to time.

(e) Policies on such factors as national defence, conservation, ownership, etc, with respect to energy.

These rules would need to be defined with due regard for our international commitments on energy, for instance as a member of the European Community.

23 If ground rules were to be adopted by each energy industry and accepted by other interested parties, not only would the Government's task of coordination be eased but a basis for monitoring general progress would emerge.

24 To sum up: such ground rules would:

(a) form a framework essential to the proper division of duties between Government and the energy industries;

(b) help in decentralising decision-taking, and and hence the genuine development of industrial democracy, by assisting the devolution of decisions to regions and functions within each industry;

(c) allow society to understand and approve the framework for long term energy planning as the only sound basis, after taking account of all interests.

The task of defining these rules within international commitments is primarily for the Government after consultation.

25 There is a need for agreement, among participants in this Conference, that once the Government has set the rules, they will be adopted generally as a settled basis for UK energy development in the long-term.

Oil

The role of the oil industry in UK energy supply and demand 1975-2000 (Petroleum Industry Advisory Committee)

1 The purpose of this brief paper is to provide a perspective on the role and significance of the oil industry in meeting the UK's widely varied demands for energy during the last quarter of the twentieth century.

2 Oil's place in the 1975 energy balance was:

	Millions tonnes of coal equivalent (mtce)		
	Total Primary Fuels	*Supplied by Oil*	*Oil as % of Total*
Markets Specific to Oil	68	68	100
Electricity Generation	114	22	19
Industrial, Commercial & Residential and Own Use	165	65	39
Total all uses*	347	155	45

*To simplify the presentation but in order to cover the total demand for energy resources the non-energy uses such as petrochemical feedstocks have been included with the energy uses.

Most oil and a small quantity of coal and natural gas were imported. Sixty eight mtce of oil went into markets where no other fuel can provide a practical economic substitute, for example transport, petrochemical feedstocks, bitumen and lubricating oils. Including the oil industry's own use, 78 mtce or over 50 per cent of oil went into oil specific end uses. The remaining 77 mtce of oil was sold in other markets where it was the preferred fuel on grounds of availability, quality, convenience or price.

3 Looking to the future there is obviously considerable uncertainty. However, making two assumptions for economic growth, $2\frac{1}{2}$ per cent pa and 4 per cent pa gives the following picture for energy demand:

Total Primary Fuels—Million tonnes coal equivalent							
	1975	*1980*		*1990*		*2000*	
		Low	*High*	*Low*	*High*	*Low*	*High*
Markets Specific to Oil	68	75	83	92	105	107	133
Fuel Competitive Markets;							
Electricity Generation	114	119	131	148	187	187	272
Industrial, Commercial & Residential and Own Use	165	173	189	201	236	240	326
Total all uses	347	367	403	441	528	534	731

The 2,000 projections are really sighting projections designed to illustrate the impact of variations in economic growth on energy demand.

However, it must be emphasised that in addition to the figures shown for oil under the above heading 'Markets specific to oil', there will be a continuing high level demand for oil in the other markets shown above even if coal or nuclear show substantial growth.

4 Taking the Department of Energy's *Brown Book (Development of the oil and gas resources of the United Kingdom 1976)* figures as a guide, annual North Sea oil production should be of the order of 170–255 mtce during the 1980s and natural gas (including supplies from the Norwegian sector of Frigg) should be around 90 mtce.

The essential change from 1975 on the supply side in meeting the demands on energy resources set out in the Table under (3) above is the advent of North Sea oil which, by 1980, could represent about 45 per cent of the UK's total demand for energy resources for both energy and non-energy uses and increase further in importance during the 1980s. Including natural gas, the proportion is about 65 per cent in 1980 with again increasing importance during the 1980s.

The extent to which the level of production foreseen in the 1980s can be sustained thereafter is very uncertain. This will be influenced not only by the level of recoverable reserves from existing concessions but also from the results of exploration in areas not yet licensed.

5 The impact of North Sea oil and gas on the economy and, in particular, the relief to the balance of payments, will be considerable and should allow the UK economy to grow at above its historical rate during the 1980s. This growth will help to provide the resources for the investment into alternative energy supplies which will probably be needed by the end of the century.

6 Given coal and nuclear's present position; long lead times for their expansion; and the likely production plateau for gas as indicated in the

Brown Book, the oil industry's flexibility (both in total volume and the range of products needed to meet different levels of demand) will be crucial in the next 10/15 years in supplying a large part of the incremental energy demand inherent in good economic growth. This is apart from the substantial growth expected in the markets specific to oil.

7 Only if the coal and nuclear industries in total show a rising contribution over the years will the advent of North Sea oil permit the UK in the 1980s to be a substantial force in international trading in oil. This would be both as a net exporter and in the optimisation of the contribution to the balance of payments by judicious importing of appropriate crudes to match UK demand while at the same time exporting equivalent quantities of North Sea oil.

8 The oil industry has already made very heavy investments in facilities to meet growth in all phases of the industry, ie not only exploration and production but also refining and distribution facilities, etc, and is therefore well placed to deal with an increase in demand. Future investment downstream will naturally be geared to demand growth.

9 The UK oil industry believes in the desirability of an energy strategy and welcomes the idea of consultation and taking part in formulation of such a strategy. In conditions of uncertainty, however, it will be important not to formulate too rigid a strategy as it is impossible to forecast precisely both total demand and the contributions of the various fuels.

10 North Sea oil will be available for a substantial period during which there will be time to work sensibly towards the likely necessary long-term developments of coal, nuclear and other energy resources without the mistakes and traumas of crash programmes.

11 The oil industry will play an increasingly important role in meeting the nation's energy requirements if sustainable economic growth is to be achieved. It is clear that oil industry operations in the North Sea make that a realistic prospect. This underlines the need for the oil industry to be encouraged to operate on a financially sound and profitable basis in both its upstream and downstream operations in order to meet the complex problems of supply and demand that it undoubtedly will have to face.

A note from the British National Oil Corporation

1 In considering the United Kingdom's future energy policy, it is as well to remember that in the post-war years many of the confident assessments of the future—whether directed to trends of consumption or the relative attractiveness of different primary fuels—have not been borne out by events. But while this experience cautions against any approach which is dogmatic, there is a better appreciation of the general outlook than heretofore. There is a clear distinction between immediate availability of primary energy fuels, and the longer-term demand and supply situation.

2 In the post-war years, primary energy demand in developed countries has been increasingly met by oil and natural gas. These two fuels are convenient and versatile, and oil in particular is easily transportable world-wide. They have displaced coal in traditional coal end uses, even when the latter is competitive in terms of price. Natural gas also replaces electricity where the latter is used in heating applications.

Oil and natural gas have been both plentiful and cheap. In the 1960s, oil became exceptionally cheap. Even after the recent and very steep price increases, oil is still not exceptionally dear.

3 Fossil fuel resources are finite. Oil and gas reserves are now being measured in decades, even if coal is still referred to in centuries. Nuclear energy, in the earlier post-war assessments, seemed to offer the world a virtually unlimited energy source. The sequence was to be fission reactors, breeder reactors, and finally fusion reactors. This is still the scenario, but the time scale and expenses of development have both escalated. There has also been increased concern about the short-term and longer-term safety aspects of a big nuclear energy programme.

Nevertheless, the twenty-first century still seems likely to be the age of nuclear power. It is possible that benign renewable energy resources will be developed, but it is difficult to see this happening on a scale capable of satisfying the world's needs.

4 The United Kingdom was for two centuries a coal-based economy, with exports of coal also playing a vital part in the country's growth and prosperity. The progressive cut-backs in coal production post-war led to oil overtaking coal as an energy source at the beginning of the 1970s. However, the falling consumption curve of the one, and the rising consumption curve of the other, have since levelled out. The most interesting development in recent years has been the growth of natural gas from the North Sea as a source of UK primary energy, a four-fold increase in gas-supplied therms bringing natural gas's contribution in 1975 to some 40 per cent of the coal and oil levels.

5 The North Sea discovery of natural gas in the sixties was of immense significance for the country. Development of North Sea oil in the second half of the seventies promises to be of even greater economic significance.

6 At a time when most developed countries are contemplating with some anxiety their position with regard to energy supplies in the 1980s, because of their necessity to import, the UK can look forward to two or more decades of energy self-sufficiency, with a possibility of establishing net exports in addition. Economic growth and industrial restructuring will not be at risk because of energy shortages, nor from balance of payment restrictions. The experience of Holland since the development of the Groningen gas reserves is an encouraging portent.

7 The current position is that UK coal supplies, under existing expansion plans, are sufficient at 150,000,000 tons per year to meet foreseen demand until 1990. The proving of workable deposits which is currently in hand, offers the prospect of increased coal supplies thereafter.

The North Sea oil fields currently under development should see self-sufficiency in oil reached by 1980, at about 100,000,000 tons per year. This figure is sustainable for a few years under existing projects, but additional reserves need to be developed to maintain this figure from the mid-eighties, or to increase it to give an exportable surplus. About half of the recoverable oil thought to be in the United Kingdom Continental Shelf is still to be identified. Two-thirds or more of this recoverable total, known deposits, and still-to-be-discovered deposits, remain to be developed.

Natural gas production is due to increase by 60 per cent in the next three or four years and could well double by the early eighties, to give 70 million tons or so of oil equivalent. And it seems likely that this level of availability could last well into the nineties, and possibly beyond.

Nuclear energy supplies will receive a boost from the Advanced Gas-Cooled Reactor (AGR) stations in the pipeline, and from the planned SGHWR units. Availability will then meet demand until 1990 or thereabouts. The long proving time for nuclear stations suggests that a commercial fast reactor prototype should be put in hand soon—this year—with the objective of getting such stations established by the mid 1990s. Success would be the key to ensuring electricity requirements for the future.

8 Coal is likely to be under pressure from natural gas in the next ten years, since the latter can take both coal's direct markets and the markets of coal's major customer, electricity. Yet for the sake of the future it is essential to maintain a present substantial position for coal. It is for consideration how to harmonise the need for this with the release of increasing amounts of natural gas to the domestic market. The Gas Industry itself sees coal as a future source of substitute natural gas, so the two industries have a similar long-term objective.

9 Oil is indispensable for transport of all kinds. It is well-suited to some fuel uses and is the most versatile feed stock for chemicals of all kinds, including synthetic fibres. It is the source for lubricants, and for a wide range of specialist products. The natural gas liquids—ethane, propane and butane—are also available for some of these fields, and natural gas itself can be used as a chemical source material.

Coal can be used as a primary material for the manufacture of all the end products derivable from oil, but the processes are expensive. They will probably be developed as oil and gas become scarce but, except in very special circumstances, not before.

10 The energy resource most likely to give the United Kingdom a net export capability is oil. It is easier and cheaper to transport overseas than gas, and is more attractive to overseas customers than coal. Nuclear energy is for home consumption, except for minor electricity load-sharing arrangements with neighbouring countries.

A detailed knowledge of the oil resources of the United Kingdom Continental Shelf is necessary in order that the energy policy for the rest of the century can be well-judged. The recoverable oil from a deposit depends on many factors, and these include number and spacing of production wells, injection techniques and number of injection points, number of platforms and their spacing, the development of subsea completion techniques, and the rate of depletion from the start of production. A full analysis can bring out the fact that the interests of the State, both from the point of view of ultimate recovery targets, and from revenue considerations, are not necessarily identical with those of private oil companies.

For maximum freedom of action, oil resources should be in the hands of the State, just as coal, gas, electricity and nuclear energy are. With North Sea oil, this is not possible. The country has an existing situation where the development of a crucial natural resource is largely in the hands of multi-national corporations, most of them foreign-owned. The multi-national corporations are likely to play a prominent part in the future development of the oil resources, although with the formation of a national oil corporation there will be increasingly a direct role for the State. It is most important that a satisfactory relationship is worked out with the existing licensees, to ensure that the legitimate interests of both the nation and the companies are met. The skill and resources which the oil companies have brought to North Sea oil exploration and development have been of the greatest value to the United Kingdom. It is these skills and resources which have made possible the opening up of some of the most difficult oil territory in the world, and they will be needed and valued for the future.

It is the Government's stated intention that BNOC—established on January 1, 1976—should develop the full capabilities of a large integrated oil company. The objective is to provide expertise for the future, with a wide applicability. The objective is achievable, as the example of other countries has shown. Progress will be quicker the more successful the participation and partnership arrangements currently being discussed with existing licensees in the North Sea.

11 The development of the offshore fields has already generated considerable industrial activity in the United Kingdom. The circumstances of the offshore business—with the expertise originally in

the hand of overseas suppliers—led to the Offshore Supplies Office being located in the Department of Energy. In this respect, energy policy has had to look wider than the fuels themselves and bring into the picture British manufacturing and design interests. This aspect will continue to be of importance as development of the fields continues.

12 To summarise: the discovery and development of North Sea oil resources can give the UK economy a much needed freedom of manoeuvre, without balance of payment restrictions, while the industrial base of the country is modernised and extended. The full assessment of these resources is a priority task. The development of the reserves will be highly capital intensive, and the timing of development will need issues of some complexity to be settled. The scale of finance and technical expertise required, and the shortness of the time scale, make it desirable that both assessment and development are carried out largely in partnership with the long established oil multi-nationals, and this is the immediate role seen for BNOC. The presence of BNOC in all North Sea oil fields being developed, and to be developed, will give the State a knowledge, and an effective voice, based on partnership, which will make easier the integration of oil production into an overall energy strategy.

UK offshore petroleum exploration and production (United Kingdom Offshore Operators' Association)

1 INTRODUCTION AND SUMMARY

UKOOA comprises 39 oil companies which are operators of UK offshore production licences.* The following paper seeks broadly to reflect the views of the UK offshore petroleum exploration and production industry, bearing in mind that UKOOA has no authority to bind its individual members to any given position.

The paper highlights the achievements of the oil industry in discovering and developing Britain's offshore petroleum resources. It shows that in the North Sea there is probably more scope for the oil industry's expertise and capability, and a proportionately greater need for risk capital, than in any other active theatre of petroleum exploration and production operations. For the nation, the benefits from the full, efficient and cost-effective realisation of Britain's offshore petroleum resources by the oil industry will remain undiminished. More particularly, the premium on technical capability and cost effectiveness will increase as production from the first generation of larger and more prolific fields begin to decline. Given this, it can be seriously questioned whether the national interest is best served by the apparent shift of emphasis in Government policy from stimulus and encouragement of the UK offshore petroluem industry to tightly-regulated control, as reflected in recent legislation.

However, the paper concludes that the requisite stimulus and encouragement to enterprise, efficiency and technological innovation need not be incompatible with the formulation and attainment of a stable, long-term national energy policy. Because of the unprecedented scale of risk investment required to develop Britain's offshore petroleum resources, neither the licensees nor the nation can afford the under-utilisation of developed production capacity which would result from imposed production cut-backs. For this reason, if a national petroleum depletion policy is adopted, it is vital that its implementation should be achieved through the controlled timing of new development projects, and likewise licence awards, against the background of a stable, long-term national energy plan.

Thus, for offshore licensees the key issue is not the need for overall planning and control at the national level, but the manner and spirit in which control is exercised. As regards Britain's offshore petroleum resources, it is crucially important that successful, efficient and innovative exploration and development should be positively encouraged and not stifled in a climate of restrictive Government control.

2 OUTLINE REVIEW OF THE UK OFFSHORE PETROLEUM INDUSTRY

Since the first UK offshore production licences were issued in 1964, the UK sector of the North Sea has been the focus of unparalleled effort and achievement on the part of the offshore petroleum industry:

(a) Some 600 exploration and appraisal wells (and over 250 production wells) have been drilled in UK offshore waters.

(b) The success of this exploration and appraisal effort is demonstrated by the UK oil and gas reserves estimates for existing licences, recently compiled by the Department of Energy in the *Brown Book,* ie 2,300 million tonnes (17 billion barrels) of recoverable oil and 40 trillion (10^{12}) cubic feet of recoverable gas, classed as proven or probable.

(c) Six gas fields have been developed in the southern North Sea. In 1975 these fields provided 97 per cent of UK gas supplies and about 17 per cent of the total supply of UK primary energy, representing the oil equivalent of over 30 million tonnes (600,000 barrels/day).

Additional gas production from the Frigg and Brent fields is under development, and is expected to raise the contribution of offshore

*The Appendix to this paper gives an outline of the composition and objectives of the UK Offshore Operators' Association Limited (p.53)

natural gas to about 20 per cent of total primary energy in 1980.

(d) Fourteen oilfields are under development in the central and northern waters of the UK North Sea. Of these three are already in production, and by 1980 the expected level of production of some 100 million tonnes per year should make the UK self-sufficient in oil, representing over 40 per cent of total primary energy.

(e) In the northern North Sea, where the majority of the oilfields are located, the combination of deep water (400 to 500 feet) with extreme wind and wave conditions has imposed more exacting engineering design criteria than any yet encountered offshore by the oil industry.

(f) Commensurately, the costs of developing North Sea oil are uniquely high; the level of investment per unit of production capacity is upwards of ten times greater than that for oil from the principal OPEC areas. In 1976 alone, the UK offshore petroleum industry will be spending some £1,500 million. Over the five years from 1976–1980 it has been estimated that of the order of £10,000 million will be invested in developing UK offshore oil and gas production capacity.

(g) From the outset UKOOA members have worked closely with the Government to establish and attain the highest environmental and safety standards. In addition, UKOOA has pioneered far-reaching initiatives, such as the Offshore Pollution Liability Agreement.

It is important to recognise that the very significant economic benefits conferred by the finding and development of Britain's offshore oil and gas resources have been realised through a policy pursued by successive Governments of stimulating the exploration and development of UK oil and gas resources within a framework of sympathetic and restrained administrative control. This in turn has encouraged the application of highly-skilled effort, technological innovation and risk capital to an unprecedented degree.

With the Oil Taxation Act, 1975, the Petroleum and Submarine Pipelines Act, 1975 and parts of the Energy Bill now in passage through Parliament, a new and much more extensive framework of legislative controls is being erected around all aspects of the UK offshore petroleum industry. These new controls, further compounded by the monitoring role envisaged for the British National Oil Corporation under the Government's objective of State participation in offshore production licences, are seen by licensees as reflecting a disturbing shift of emphasis from stimulus to restriction. The licensees' consequent concern does not stem from belief that their interests are in any way incompatible with the national interest, in furtherance of which the new controls will be applied. Rather, it stems from serious concern at the potentially stultifying effect of over-regulation This is increased by the current lack of progress being made in translating legislative principles into practicable administrative effect.

3 NATIONAL INTEREST AND THE COMMERCIAL INTERESTS OF THE UK OFFSHORE PETROLEUM INDUSTRY

With regard to the development of Britain's offshore petroleum resources, there has been a common tendency to believe that the commercial interests of the production licensees are in direct conflict with the national interest. It is suggested this need not be so and that the conditions which are conducive to high-risk commercial enterprise are compatible with the wider national interest. The ultimate quantities of oil and gas which can be produced commercially from a given hydrocarbon province depend to a significant extent on the quality of the effort applied to finding and developing the reserves. This is particularly true of the North Sea, where the uniquely high costs of petroleum exploration and production mean that there is a correspondingly high proportion of fields (large and prolific by normal standards) which are marginally commercial; hence the considerable incentive for technical innovation and ingenuity.

Hitherto the policy of successive British Governments of stimulating rapid exploration and development of the UK Continental Shelf has induced the oil companies to dedicate substantial resources and their most capable staff to this venture. Experience and talent drawn from many theatres of operation have been directed to opening up the UK North Sea and to pioneering new techniques for developing its potential. The benefits of this are strikingly reflected in the current UK reserves and production estimates.

Now, however, the build-up of Britain's offshore petroleum resources is entering a new phase. The Government have stated that they no longer necessarily regard the national interest as being served by continued rapid development, and a more measured pace of new development activity, in compliance with a national depletion policy, is in prospect. Nonetheless, it is submitted that whether or not such a policy is adopted there will continue to be an undiminished requirement for ingenuity and expertise of the highest order if the benefits of Britain's offshore petroleum resources are to be fully realised. For example:

(a) As mentioned above, the cost of producting UK North Sea oil is several times greater than the cost of producing OPEC oil. Accordingly, there is a corresponding premium on efficiency and cost effectiveness.

(b) The maturity of UK North Sea exploration is such that exploration effort will tend to yield diminishing returns. New discoveries will tend to be smaller and more marginal, creating both the need and scope for the difficult task of achieving breakthroughs in cost-saving production technology.

(e) As with any oilfield, the net recovery of oil from North Sea fields is limited, and the application of new techniques for enhancing the average recovery could offer substantial benefits.

(d) As development of the North Sea proceeds there will be a growing 'infrastructure' of established facilities such as production platforms, pipelines and tanker loading units. When these facilities cease to be fully utilised by the fields for which they were initially constructed, there will be increasing scope for using them as the basis for the development of reserves which otherwise would not be commercially recoverable. However, the duration of such opportunities will be limited to the safe, operational life of the individual components of the 'infrastructure'. Accordingly, the emphasis will be on ingenuity and innovative technology, aimed at maximum realisation of the incremental development opportunities within the limited time available.

Thus, it is in the wider national interest that the oil industry should be induced by commercial motives to continue dedicating to the UK offshore venture its most experienced and capable manpower resources, backed by risk capital and worldwide research and development.

4 THE FRAMEWORK FOR A UK PETROLEUM DEPLETION POLICY

It is recognised that the development of Britain's offshore petroleum resources is now so central to the UK economy that the Government will wish to establish and maintain a national petroleum depletion policy consistent with the wider national interest. Also recognised are the implications for employment in the UK of the Government's recently-stated objective of maintaining offshore activity at a reasonable stable level in the future.

However, there is already serious concern within the industry that the nature of the controls contained in the large volume of recent legislation may have a stultifying effect on UK offshore exploration and production activity; likewise, the manner in which a national petroleum depletion policy is implemented. It is feared that the Government's understandable desire for the necessary degree of control may lead to excessive regulation, so that ultimately the effect is counter-productive and contrary to the national interest, with the result that:

(a) commercial enterprise is suppressed and risk investment on the massive scale required is inhibited,

(b) the incentive for technological innovation and ingenuity is stifled,

(c) the reward for efficiency and cost effectiveness is removed.

It is of paramount importance that Government controls over the UK offshore petroleum industry should be applied in a manner which fully recognises the benefits to the national interest which flow from encouraging the private sector.

More specifically, because of the enormous investment entailed in developing UK offshore oilfields, it is essential that the depletion controls should be such as to ensure that once fields are developed they are utilised at their optimum capability. For this reason, the primary instrument of the national oil depletion policy should be the co-ordinated timing of the issuance of licences and any subsequent development programmes against the target of a stable, long-term plan of the desired national oil production profile. Controlling the level of national production through the cutting back of developed oil production capacity should only be regarded as a short-term measure for fine-tuning in the last resort.

The essence of the offshore petroleum industry's requirement is the establishment and consistent pursuit of a stable, long-term UK energy policy within which the full, efficient and cost-effective realisation of Britain's offshore oil and gas resources is encouraged. To this end, UKOOA welcomes the Government's present initiative in calling a national energy conference and looks forward to continuing consultation.

APPENDIX

Offshore Operators' Association Limited

In June 1973 the offshore operators in United Kingdom waters converted their earlier organisation—The North Sea (UK) Operators Committee, formed 1964/65—into the present UK Offshore Operators Association Limited, being a registered company limited by guarantee and having no share capital. There are 39 members.

In addition to periodical general meetings, there are monthly meetings of the Council of 20 representatives, elected from member companies.

The basic objectives of the Association are:

1 To provide an industry forum for discussion of technical and administrative matters common to the industry relating to exploring and producing in all designated waters on the UK Continental Shelf.

2 To the extent appropriate and practicable in various circumstances, to provide for the industry—on technical and administrative matters of concern to members—a means of communication with the government and others.

The petroleum industry is multi-national. It has been mainly in the USA in some European countries and in the Middle East that the industry has acquired the experience, skills and specialised techniques of exploration and it has been largely in those territories that the hundreds of millions of pounds devoted to the search for oil and gas have been raised and spent. The rapid deployment here of expertise acquired abroad has chiefly brought about the establishment in these islands, within a single decade, of highly important petroleum prospects. (It was, however, a British company that made the first commercial strikes of gas and of oil.)

It is natural, therefore, that the interest behind the membership of UKOOA are now predominantly foreign. Though all holders of production licences on the British Continental Shelf are companies registered in the United Kingdom, it will be understood that those whose 'parents' belong to other countries tend to have different circumstantial backgrounds and juridicial circumstances, rendering it difficult for UKOOA to answer all question or speak for every member under all conditions.

Officers

ELECTED

President

W E Bell
Shell UK Exploration & Production
Shell Centre
London SE1 7NA
01-934 1234

Vice President (England)

E A Freeburg
Arco Oil Producing Inc
Collier House
163–169 Brompton Road
London SW3 1PZ
01-584 0581

Vice President (Scotland)

M M Linning
BP Petroleum Development Limited
Farburn Industrial Estate
Dyce
Aberdeen AB2 0PB
0224 723131

Hon Secretary

W A Sax
Unionoil Company of Great Britain
32 Cadbury Road
Sunbury-on-Thames
Middlesex
76 85600

Hon Treasurer

P A Taylor
Hydrocarbons Great Britain Limited
59 Bryanston Street
London W1A 2AZ
01-723 7030

APPOINTED

Director General

G Williams
UK Offshore Operators Association Limited
192 Sloane Street
London SW1 9RB
01-235 0762

Executive Secretary

A E C Staines
UK Offshore Operators Association Limited
192 Sloane Street
London SW1 9RB
01-235 0762

Committee Officers

Chairman	Secretary
Aberdeen Emergency Action Group	
R. J. Ball Amoco (UK) Exploration Company Aberdeen	R J Salt Conoco North Sea Inc Dundee
Advisory Committee on Offshore Protection	
G. Williams UK Offshore Operators Association Ltd	Group Captain A Smart Shell UK Limited
Aircraft & Communications Committee	
J Dorman Mobil North Sea Limited	M G Cherry Conoco North Sea Inc
Clean Seas and Emergency Services Committee	
R J Fulleylove BP Research Centre	D R Blaikley Amoco Europe Inc
Employment Practices Committee	
J A G Cotton BP Oil Limited	D A Cuthbertson Chevron Petroleum Co Ltd
Engineering & Development Committee	
J Dear North Sea Sun Oil Co Ltd	G N Rumbaugh North Sea Sun Oil Co Ltd
Exploration Committee	
M Ramsden Conoco North Sea Inc	E H East Unionoil Company of Great Britain
Finance Advisory Committee	
C A E T Stevenson British Petroleum Co Ltd	E J Milan Mobil Producing North West Europe Inc

Committee Officers

Chairman	Secretary
Full & Fair Opportunity Committee	
Co-Chairman E J Hannington BP Trading Ltd	Co-Chairman D W Jones Esso Exploration & Production UK
Gt Yarmouth Emergency Action Group	
R J T Briggs Shell UK Exploration & Production Lowestoft Base	R Martin Conoco North Sea Inc Great Yarmouth
Information & Public Affairs Committee	
R Dyk Hamilton Brothers Oil & Gas Ltd	M J Wells Amoco Europe Inc
Interest Relief Grants Committee	
J Badger Shell International	
Marine Committee	
Captain L G Buckenham BP Petroleum Development Ltd	H A Twiddy Total Oil Marine Ltd
Medical Advisory Committee	
Dr R Houston Shell International Petroleum Co Ltd	Dr C Jones BP Petroleum Development Ltd
Non-Fiscal Legislation Committee	
R Dyk Hamilton Brothers Oil & Gas Ltd	J B Wybrew Shell UK Exploration & Production
Oceanographic Committee	
T N Watson Amoco (UK) Exploration Company	
Offshore Fire & Emergency Control Committee	
E G Everett Shell UK Exploration & Production	No Secretary
Safe Drilling Practices Committee	
R D Koch British Petroleum	No Secretary
Safety Committee	
C Allen BP Petroleum Development Ltd	J E Howard Shell UK Exploration & Production

Nuclear

The future of nuclear power in the UK (United Kingdom Atomic Energy Authority)

The need for electricity

1 Economic growth and the opportunities it offers for enhancing the material and social well-being of the community will lead to greater need for electrical power. If the well-being of the community is to be maintained and enhanced by increased industrial production, higher standards for social and other services, larger disposable family incomes and employment opportunities, there is strong evidence that this creates additional demands for energy in the form of electricity. Such demands are likely to go on growing up to and beyond the end of the century.

The need for nuclear power

2 The predicted future scarcity of oil and gas at economic prices, and the consequent need for coal to act as a source material of carbon to replace them—both as fuels and as chemical raw materials—combined with these needs for electricity, will put great pressure on future fuel sources. This will bring a real and increasing risk that we could no longer afford the fuel required to maintain and improve our living standards. Nuclear energy is the only new and additional energy source that can be made ready in time and in sufficient quantity to meet the stiuation. It is therefore essential to use the alternative of uranium to augment the other sources of energy in order to meet the need for additional low cost supplies of electricity.

3 Fortunately, a steady build-up of nuclear stations for generating base-load electricity can take place without economic penalty, as can be seen for example, from the 1974/75 generating costs for modern Central Electricity Generating Board (CEGB) stations. Similar evidence comes from other countries. The economics of nuclear electricity, like that of coal, will depend in part on the level of activity in the industry and, notwithstanding the good long-term prospects that nuclear power shares with coal, on the short and medium term prospects which face the staff who work in it.

4 Although uranium prices will rise as uranium in limited supply like other fuels, becomes scarce, nevertheless given moderate stability, nuclear power can continue to be economically attractive; even the present generation of thermal nuclear reactors, which use only about one per cent of the energy in uranium, give a heat output from each ton of uranium equivalent to about 20,000 tons of coal. Thus despite a rising price trend the low fuelling costs of uranium will continue fully to offset the additional capital costs associated with nuclear power stations and the fuel supply and treatment plants required to service them.

5 Eventual scarcity of uranium itself does not inhibit the use of nuclear power because another class of reactor, 'fast breeder' reactors, can be used to extract 50 per cent or more of the energy in uranium, taking from each ton energy equivalent to 1,000,000 tons of coal. These reactors can use the uranium from which the thermal nuclear stations have taken their 1 per cent. Large prototype fast reactors are already operating here, as well as in France and Russia.

6 The use of fast reactors offers then a huge source of energy for use at the time—around the end of the century—when economically extractable fossil fuels and uranium become short relative to demand. But we cannot afford to wait until then to demonstrate and prove the use of these reactors. Fast reactor power stations will, certainly in the early years, be more expensive to build than other types; but again the higher capital costs are expected to be offset—and eventually more than offset—by the lower costs of fuelling the stations.

7 Because of the history of its inception, nuclear power understandably evokes public concern over environmental issues. But largely because of the recognition from the outset of the need to apply the most stringent safety standards, the use of nuclear power has had an excellent safety record to date and caused no significant damage to the environment, and has some

environmental advantages over its alternatives. With the progressive improvement in safety precautions in reactors, even beyond today's standards, and the introduction of improved fuel processing methods and provision for long-term waste storage, there is no reason to doubt that this already good safety and environmental record will be maintained.

Resources required

8 Although the use of nuclear reactors to generate cheap base-load electricity has been successfully established in the UK and other countries, the nuclear and engineering industries have big tasks ahead. If nuclear power is to be maintained as a means of meeting base-load electricity needs of the future in the most economical, as well as safe and environmentally acceptable, manner, then in addition to the operating teams in the Generating Boards, there are needed:

(a) Enough skilled men to engage in research and development, in reactor design and in fuel cycle services. This condition should be met by the strong alliance now established of three organisations with clear responsibilities for each of these functions:

The United Kingdom Atomic Energy Authority—research and development.

The Nuclear Power Company—reactor design and construction.

British Nuclear Fuels Limited—design and operation of fuel, enrichment, fabrication, reprocessing and waste storage plants and facilities.

(b) In those parts of the UK engineering industry which supply equipment for power stations and fuel cycle plants, teams of experienced, skilled workers and managers, with supporting assets, on which to build an expanding programme.

9 There are resources in the UK such that these tasks can be achieved but only if some stability is attained in the ordering pattern for power stations, to an extent consistent with the medium term expectations for electricity needs. It is apparent from the difficulties of some sections of the boiler, generating equipment, station construction and fuel supply organisations, arising from the present variable rate of ordering, that without greater stability it will not be possible to maintain viable manufacturing teams. Given this stability, then it should be practicable to:

(a) maintain nuclear power as the cheapest form of base-load electricity, with adequate assurance that it is both safe and environmentally acceptable;

(b) provide a sound basis for expansion in the supply of power stations as and when demand increases;

(c) establish, making full use of our own extensive experience, in collaboration with other countries, the operation of full-scale fast reactors so that they can confidently be used in the UK and elsewhere in the numbers that will be needed at the end of the century and beyond;

(d) provide the necessary domestic base for the exploitation of near and medium term export opportunities by the reactor component supply industry and in the provision of nuclear fuel services.

10 Other countries will shortly overtake the UK in their capacity for nuclear generation. They accord high priority, for reasons both of economy and security of supply, to having relatively cheap nuclear power available. By maintaining UK capability we can keep manufacturing industry on terms with its competitors for the economical supply of the electricity it must have. We can enjoy the various benefits particular to availability of electricity at reasonable prices. We can avoid the prospect of substantial imports of plant and machinery in the 1980s and 1990s, and eventually with the use of fast reactors we can look forward to a prolonged further period of lessened dependence on imported fuels after North Sea oil and gas run down.

Energy policy and the nuclear industry (Nuclear Power Company Limited)

The events of the last few years show that in the short term energy demand responds not only to changes in industrial and domestic prosperity but also to changes in relative price. The real price of most forms of energy has suddenly become about as high as it was in the 1950s. The exception is provided by gas, the price of which has fallen substantially in real terms over the last 20 years. As a result, at a time when the growth in total energy use has been negligible there has been a significant redistribution of existing demand in favour of gas.

These effects have taken place over a few years only, but because some of the contributory causes persist they cannot be ignored in making long-term forecasts and they introduce greater uncertainties in these forecasts than have generally been thought to be present in the past. Nevertheless, long-term forecasts, or some other basis of decision is necessary in many of the energy industries because decisions must be made to provide major plants which are intended to meet needs 10 years ahead. Indeed, if the original decision relates to a prototype or demonstration unit which must operate before the main series of plants can be ordered, the lead time may be up to 20 years.

It is reasonable to ask whether forecasts of requirements 10 to 20 years ahead can be regarded as a sufficiently reliable basis for an energy policy, to ask what are the consequences of being wrong and whether there is a basis of decision which is not so dependent upon demand forecasting. As a result it has become the practice to investigate a number of future scenarios based upon different patterns of growth and consumer preference. These studies illustrate the diversity of answers that can be obtained from different assumptions, but they do not always indicate which things must be decided now and which decisions can be delayed in the hope that the picture will become clearer in a few years' time.

What is sought is a solution which gives the greatest flexibility to respond to changes in energy demand, or preference, which minimises the need to direct the preference of consumers by selective taxation and which, at any given time, is not much more costly to the community than the solution which could have been devised if the future could have been foreseen. Obviously, it will often be impossible to devise such a solution in a complete form, but it may be possible to recognise the contribution to future needs and flexibility which could be afforded by individual decisions and to execute these decisions with fair assurance that they will not be regretted.

In some advanced countries with small coal and oil reserves, and in others that prefer or need to export fossil fuel, a large nuclear power programme appears to have such advantages that the rate of installation of nuclear capacity is limited not by preference but by financial resources. In some cases the nuclear capacity will provide not only for growth in electricity demand, but also for the large scale replacement of fossil fuelled power stations. In the UK, however, the prospective availability of natural gas, oil, coal and nuclear generating capacity is such that the scope for a flexible policy is only exceeded by the scope for debate on what that policy should be. As a contribution to the debate the following propositions are offered.

1 Only exceptionally will it be possible to decide how the real long-term costs associated with a particular source of energy should be measured and to calculate it with sufficient accuracy to allow these costs to be used as a starting point for a single strategy.

2 The UK must retain a viable coal industry with capital investment and a wages structure calculated to provide the necessary recruitment for the 1980s.

3 The UK must minimise its dependence in the long term on imported fuel because political and demand considerations will tend to increase the real costs and uncertainties of supply of such fuels.

4 If, as seems likely, a transition from natural gas to substitute natural gas is required, the pricing of

the natural material should increasingly take account of the long-run cost of the substitute material so as to avoid the development of an industrial and domestic structure which cannot be sustained.

5 Because a viable electrical power plant manufacturing and construction industry is a requirement of a flexible policy, it will be essential to ensure that this industry, restructured if necessary can survive with the relatively small volume of home business in prospect for the next several years. It may also be possible to schedule that business so as to contribute to the solution of the problem.

6 The arguments of (5) apply with equal force to the nuclear power plant design, development and construction industry which can contribute by providing a programme of thermal reactors in the 1980s and which will be necessary in the long term of the UK is to have the option of installing fast reactors and so obtaining almost perpetual independence in fuel supplies for electricity generation.

It will clearly be for each industry to evaluate the degree of flexibility and future protection afforded by a particular pattern of decisions for that industry, with less emphasis than has sometimes been the case upon increased market share as an end in itself. The remainder of this note is concerned with a viewpoint on this matter from the nuclear industry.

The short-term position is that by 1980 the UK will have in operation about 10,000 MW of nuclear power capacity which, at base load operation, will save the equivalent of 15 million tons of fuel oil per year, or £500 million per year at present prices.

The characteristic variations in the demand for power in the UK throughout the day and the year are such that a substantial part of the installed generating capacity will only be called upon to operate during most years of its life for a fairly small proportion of the time. For such capacity capital costs are proportionately more important and fuel costs less important than in base load operation, and there is no prospect that it will be nuclear-fuelled for many years.

But between this capacity and the 10,000 MW of nuclear stations which will already be in operation, there should by the later 1980s be the opportunity to deploy, with economic advantage, several thousand megawatts of further nuclear capacity. The exact load requirements from the total nuclear component of the system will depend upon growth of electrical demand during the 1980s, which cannot at present be predicted sufficiently accurately, and the direct economic advantage will similarly depend upon the movement in real terms of fossil fuel costs. But unless growth in electrical demand virtually ceases, the worst that can be asserted is that the capacity may be installed a few years before it is essential.

The advantages to be secured for this course are considerable. It will make it possible to have a modest but steady ordering programme for nuclear stations, typically 1,000 MW per year over the next several years and to ensure that the UK retains the option to design and construct nuclear power stations using its own resources. Most estimates of energy supply and demand in the UK foresee an increased role for nuclear stations in the long term and do so with sufficient certainty to indicate that the retention and improvement of the existing nuclear construction industry is most desirable. Furthermore, experience over the last decade in all those countries which have used domestic industries to provide the nuclear component of their electricity system shows increased emphasis on quality and quality assurance as a contribution to confidence in both safety and reliability. These requirements are only likely to be met if there is continuity of demand which gives the opportunity to prove the chosen nuclear system in practice. Practice in this case includes a proven and demonstrated safety case so that the system is known to have public acceptance for the construction of additional capacity as required. It includes reactor construction which allows the evolution in the light of experience of better design features and methods of erection. It includes the establishment or preservation of an industrial infrastructure accustomed to manufacture components to the standards now required by the nuclear engineering industry, and which may come to be required for other advanced engineering products. Although the benefit of 'spin-off' of high technology is an argument which, through over-use, has become discredited, the UK has suffered by comparison with other advanced countries from insufficient dissemination of high quality engineering and the advantages and satisfactions that go with it. The nuclear industry has in the past and can continue to provide one source of advancement of engineering standards in the UK.

A separate and important advantage comes from the opportunity to secure an option to adopt the fast breeder reactor in due course. The technical and social questions involved in the development of a plutonium breeder system have been well publicised. It is generally accepted that the development and demonstration of the system will require the resources of the Community rather than those of any single European country. It is not seriously challenged that the plutonium breeder reactor in one or other form will give virtual independence of other fuel supplies, including imported uranium, to the nuclear component of the system. It is not necessary to assert that the adoption of the breeder reactor is essential, although this is a widely held view amongst those who believe that growth in electrical energy demands will continue in the advanced countries. It is sufficient to accept that for the UK there is a need for the next decade to preserve an option to adopt the fast breeder reactor. Such an option requires that the UK should have the competence and the access to information to enable it to make its own judgements about performance, reliability and safety, and to satisfy the population at large that these have been thoroughly made. For this purpose, participation by the UK nuclear industry in the international development will be essential.

To allow the nation to proceed with a limited programme of thermal reactors, and to provide the capability to exercise the fast breeder option, requires the UK to maintain a domestic nuclear industry of competence and acknowledged international standing.

Energy conservation (Advisory Council on Energy Conservation)

The need for energy conservation—short term

1 The net import cost of fuel adds £3½ thousand million to the national import bill at a time when the deficit is at previously unheard of levels. Any reduction in energy usage contributes directly if oil, or indirectly if other fuel, to the balance of payments situation. The value of oil imports has risen to one-fifth of total imports since prices were increased in 1973/74, and no other basic imported commodity offers such wide-spread opportunities for savings. From the present time until North Sea oil is flowing in large quantities a substantial proportion of the national income must be allocated to pay for oil imports at a time when the requirements of international finance may inhibit planned growth in the economy. Energy saving now by reducing the deficit in the balance of payments will allow greater flexibility for government action.

The need for energy conservation—medium term

2 There is a tendency to regard oil and gas from the North Sea as some sort of private supply of free energy which will relieve us of the need to pay for our energy. It will be of great economic value to the country but only if we obtain the maximum price. The more that can be exported directly or converted to chemicals or other oil-based products the greater the contribution to the balance of payments and to the repayment of accrued debts. All this leads to the need for energy conservation in the medium term so that our indigenous fuel is used as economically as possible and we reap the maximum advantage from it.

The need for energy conservation—long term

3 North Sea resources are limited. No one can be sure when they will run out—but at some time in the future alternative sources or alternative types of energy will be needed. Other papers at this Conference are considering this problem under various headings and it is not the purpose of this paper to argue whether an energy gap will occur in the 1990s or at some later time. It is reasonably clear that alternative sources of energy are unlikely to be available on a significant scale in the 1990s and that proven reserves of fossil fuels other than coal are insufficient to satisfy internal requirements beyond the 1990s. Until further reserves are discovered there must remain the grave danger of an 'energy gap' occurring after a period in which we have become accustomed to self-sufficiency in energy. The impact at that time on balance of payments and the reversion to recent experience of economic growth being constantly retarded by financial problems is possibly the lesser danger. As world supplies may also be in increasingly short supply it might not even be possible to maintain the desired energy supply. Until reserve levels of oil and gas promise a higher margin of safety the need for energy conservation to delay the energy gap remains prudent.

The benefits from energy conservation

4 Energy is a basic resource in almost every industry but only in a few industries is the use of energy large enough to be a dominant cost in the industrial process. As a result there has been a tendency to treat energy as an overhead not subject to direct cost control. The stimulus to energy saving provided by the sharp price increases in 1974 caused larger industrial users to take prompt action. Substantial savings have been achieved by industrial concerns such as the Coal Board, Courtaulds, ICI, British Leyland, EMI, Marks and Spencer, Tesco, Esso and other large users of energy sometimes as a result of quite simple measures. But smaller firms do not have the necessary resources or expertise to allocate to energy audits and it is believed that there are still large savings in energy possible

at economic investment rates. Similarly the large property institutions have taken the necessary steps to achieve energy savings economically. For example the Public Services Authority of the Department of the Environment by investing £5 million p.a., made savings of £8½ million p.a. and hopes eventually to save £25 million p.a. Buildings use 45 per cent of all energy consumed in this country and there is still an immense amount of economic saving possible in the smaller commercial properties and in domestic housing. The compelling argument for energy conservation is that measures that pay for themselves on a national scale can be undertaken now to give short-term balance of payment relief and help to improve our strength in the medium term and to avert a long term energy gap.

Progress to date

5 On December 9, 1974 the Secretary of State (Mr Varley) announced a 12-point energy saving programme covering pricing policy, better insulation of buildings with special action on public buildings, speed restriction to conserve petrol, a loan scheme to encourage and assist industry, restriction on lighting and a considerable advertising and publicity campaign. It is impossible to calculate precisely the effects of the different measures but the practical effect of the sharp increase in prices of fuel coupled with the psychological effect of the 'Save It' campaign may have been the dominant contributing factors in the estimated reduction in energy consumption of 2 per cent (equivalent to £150 million pa). Other important energy conservation measures have been the restriction in maximum permitted driving speeds, improved regulations on insulation levels for new dwellings, the tax allowance to cover insulation of industrial premises and the heating order for a maximum heating level of 20°C in non-domestic buildings. In July 1975 the Select Committee on Science and Technology reported on Energy Conservation* making over 40 detailed recommendations on methods of energy conservation that could, and in the Select Committee's view should, be followed. The White Paper giving the Government's reply is expected shortly**.

Advisory Council on Energy Conservation—ACEC

6 The Council's Terms of Reference cover the identification of energy-saving opportunities, advising on energy use and stimulating the efficient use of energy. From the outset these tasks have been approached under two headings—short-term measures to assist the national economy problem brought about by the quadrupling of the price of crude oil in 1973/74 and long-term problems likely to arise from the limited nature of present sources of energy. In July 1975 ACEC made some 40 recommendations for Energy Savings in a Report to the Secretary of State*. Although the timing of the report and the number of recommendations coincided with the Select Committee report, the recommendations overlapped and produced together a combined list of some 60 methods of achieving greater savings. The ACEC recommendations were based on the Meetings of the Council and its separate Working Groups covering General Policy (Chairman, Sir Derek Ezra), Industry and Commerce (Chairman, Dr Pearce), Buildings, (Chairman, Mr Bennet), Transport (Chairman, Mr Osborn), Publicity and Education (Chairman, Mr Hammond). The Council and its Groups have consulted with 14 major industrial organisations who account for over 50 per cent of industrial energy consumption, with the TUC, and the CBI, with trade and professional groups including the major professional organisations concerned with building design techniques, with a wide variety of transport organisations and with European and International energy conservation bodies. The Council with its Working Groups therefore provides a forum for the discussion and interchange of ideas on the better use of energy which has stimulated activities in both the public and private sectors. As a result of its contacts with many different users of energy, it has seen a growing and widespread interest in energy conservation which has already had an effect in industry and commerce and in the domestic and public sectors. It is expected that the efficiency of energy use will continue to improve as the effects of new designs and techniques become apparent.

Priorities for energy conservation

7 There are now a large number of different energy conservation measures being discussed. These include technical measures such as heat and electricity generation, improved vehicle design, long distance transmission of heat, heat pumps, fuel cells and solar energy. At the moment the amount being spent on these measures is not great but if the possibility of an energy gap in the 1990s becomes stronger then it is evident that much larger sums will need to be spent on such measures

**First Report from The Select Committee on Science and Technology: Energy Conservation* (HMSO 1975)
**Now published by HMSO.
Energy Conservation Cmnd.6575

*Energy Paper Number 3 (HMSO 1975)

At the moment many longer term energy conservation schemes appear to be uneconomic, even though they may save appreciable amounts of energy. Examination shows that their unattractiveness depends greatly on the assumptions made about the trends in fuel prices and on discount rates. The measures with large potential savings may be obscured by many measures that though worthwhile offer by comparison much smaller benefits. The priorities for energy conservation policy must be to progress actively those measures that offer the largest potential savings such as those in industry and in buildings while allowing other measures to continue without diverting attention from the primary areas where large savings can be achieved.

Buildings

8 The programme of extending thermal insulation requirements in building regulations should be pursued as fast as resources and consultation procedures permit. Structural insulation requirements should be extended to cover all types of new buildings, minimum insulation standards should be established for hot water storage tanks, boilers and pipe work and wherever possible measures that increase the thermal efficiency of existing buildings as well as new buildings should be pursued, in particular methods of controlling heating equipment.

Industry

9 ACEC recommendations include that priority is given to the setting up of an Advisory Service to make available expertise on energy conservation particularly for small and medium sized firms: that companies should be encouraged to record for their own benefit energy use by source, total energy per unit of product/or raw material and an energy impact statement on these lines should be required from any industry receiving government financial assistance: that the Loan Scheme for Industry has not been widely used and a revision of terms may be necessary to provide an incentive for industry.

Conclusions

10 One of the most satisfying features of the Council's first 18 months in existence has been the increased recognition by industry and the general public of the need for energy conservation. This paper emphasises that this is not just a temporary need but that energy conservation is here to stay. The methods which will be used can be chosen from the following options:

(a) Exhortation ie publicity, information, advice training, education and the stimulation of interest in the better use of energy.

(b) The price mechanism of the 'correct' pricing of fuel and power.

(c) Fiscal incentives ie taxes, tax relief, loans and grants.

(d) Regulations and Standards, ie speed limits, vehicle economy standards, lighting, heating and air conditioning limitations, regulations or standards for building insulation and ventilation, for heating and control equipment and more generally for all energy using appliances.

(e) Action by communal authorities and industries, eg district heating, combined electricity and heat and steam generation, public transport and urban planning.

(f) Controls ie allocations, import controls, rationing.

(g) Stimulation of research and development on energy conservation and the use of renewable energy resources.

(h) Energy may be conserved not only by improving the efficiency with which it is used but also by adopting a different or more frugal way of life. Fuel shortages may greatly influence future life styles just as future life styles may have an impact on energy use.

These are the options that all of the community will have to consider in the formulation of energy conservation policy.

Research and development

Energy R & D in the United Kingdom: a discussion document (Advisory Council on Research and Development)

The following is a summary of the discussion document on energy research and development in the United Kingdom which was issued on June 8, 1976 by the Secretary of State for Energy's Advisory Council on Research and Development for Fuel and Power (ACORD).

This document contributed essential background to the National Energy Conference and is being published separately in the Energy Paper Series of the Department of Energy (Energy Paper II, HMSO).

The Advisory Council on Research and Development for Fuel and Power (ACORD) was set up in 1960. Its terms of reference are:

'To advise the Secretary of State on the general programme of research and development of the nationalised energy industries; and to advise on such other matters in the field of energy research and development as the Secretary of State may refer to it.'

The members of the Advisory Council are: Dr. Walter Marshall, CBE FRS (chairman), Chief Scientist, Department of Energy; Dr. P. J. Agius, Esso Petroleum Company Limited; Mr. P. Blakeley, Imperial Chemical Industries; Sir Sam Edwards, FRS, Science Research Council; Mr. L. Grainger, National Coal Board; Dr. T. N. Marsham, OBE, UKAEA Reactor Group; Mr. L. J. Mills, National Coal Board; Mr. M. M. Pennell, CBE, British Petroleum Company Ltd; Mr. R. W. Orson, The Electricity Council; Mr. D. E. Rooke, CBE, British Gas Corporation; Dr. H. Rose, International Research and Development Company Ltd; Dr. D. Littler, Central Electricity Generating Board.

The purpose of the discussion document prepared by ACORD is to provide a framework for the planning of UK R&D in energy technologies. It is seen as the first step towards the formulation of a national energy R&D strategy.

In a foreword, Dr. Walter Marshall, chairman of ACORD, says that a detailed strategy can be evolved only through discussion among the public and private energy industries and energy consumers. He offers ACORD's ideas in the document for constructive public comment.

The objective of the studies and discussions will be to formulate a strategy sufficiently robust to cope with the uncertainty of the future.

The framework put forward in the document:

—identifies those technologies which would keep open options over a wide range of possible future conditions with the objective, if possible, of obtaining broad national consent to the importance of pursuing them;

—gives priority to technologies which ensure that the UK derives maximum advantage from its indigenous reserves of North Sea oil and gas;

—gives priority to energy conservation, coal and nuclear technologies because they will be essential to the UK over a wide range of possible future conditions;

—seeks to use the breathing space given to the UK by North Sea oil and gas to assess thoroughly the alternative technologies and to invest in developing those selected for opening valuable options at reasonable cost; and

—seeks every opportunity for international collaboration because the problems are international.

Chairman's foreword

In his foreword, Dr. Marshall says that the UK has a favourable energy outlook compared with most of its major trading partners. But there is a consensus that, perhaps not until the turn of the century, there will be a widening gap between overall demand for energy and indigenous supplies of fossil fuels—ie coal, oil and gas.

It is therefore important to identify the contribution that energy R&D can make to bridge the gap.

Dr. Marshall then says most energy supply and demand decisions that will affect the UK in the short term have been taken and announced to Parliament. Those which will influence the next 10 years are primarily investment decisions in the nationalised industries with no great implications for R&D. But for the longer term most decisions on energy matters require a technological input and possibly considerable R&D programmes.

The dominant aim of the UK's R&D strategy must be to open technological options and a subsidiary aim must be to do as much as possible in co-operation with other countries.

Energy supply and demand

The document says that if the growth in primary energy consumption were to return to the average annual rate of nearly 2 per cent over the last 20 years consumption would double by AD 2010. North Sea oil and gas should bring important economic benefits but their production seems likely to peak 'all too soon', possibly before the end of the 1980s. Not long after that a gap would begin to develop between UK total energy demand and its indigenous supplies.

If that gap increased beyond the contribution from nuclear power, the UK would again become dependent on imported fossil fuels and would then be more vulnerable to increases in world energy prices at a time when those prices would be sharply rising as world oil resources became depleted.

'This is another reason for the UK to make sensible use of its indigenous resources and to prepare for the future both by becoming a more efficient user of energy and by developing additional sources of energy', the document says.

There is no way of being sure when the medium-term period of UK self-sufficiency will end, but it is prudent to plan UK R&D on the assumption of earlier rather than later need.

The document adds:

'There is a danger that the prospect of energy self-sufficiency, albeit for a limited period, could cause complacency . . . This would be a grave mistake, because the only major difference between the UK and countries less favourably endowed with indigenous fuel will be that, for a time, our balance of payments position will be eased.

'We must . . . take advantage of this breathing space given to us by North Sea oil to improve the efficiency of energy use and to develop, prove and establish those technologies upon which this country's future is likely to depend.'

The document examines energy supply and demand against the background of losses of 60 per cent of primary energy on its way to the consumer —30 per cent in conversion and distribution and a further 30 per cent at the point of use.

Supply

It says that *coal* has the potential to supply all UK energy needs for several centuries. Without investment, however, the UK cannot be prepared for a future in which the demand for coal will be as large or larger than it is today. It would not therefore be appropriate to judge investment with long lead times against the short-term depression in coal's traditional markets.

North Sea reserves of *oil* might rise as exploration continues, designated areas are expanded and recoverability (now estimated to be only between 30–45 per cent of the oil in place) improves.

World oil production is expected to reach its peak in AD 2000—a decade later than the North Sea's peak. But eventually the supply of oil for chemical feedstocks will be a world-wide problem, as will the supply of motor fuels unless other energy sources for transport are found.

The world-wide decline of oil supplies will probably make it essential to generate substitute natural gas from coal and to turn to coal as a supplementary source of transport fuels, organic chemicals and other premium hydrocarbons. These applications could eventually make coal too valuable to burn for bulk heat raising purposes.

After the turn of the century, the *gas* industry will be seeking alternative supplies either by manufacturing substitute natural gas or by importing liquefied natural gas.

Unless there is much more prospecting, world reserves of reasonably priced *uranium* (all of which the UK has to import) could be seriously restricted before 1990. It is therefore important to make the best possible use of uranium. It can be made to meet demand for centuries by the introduction of the fast reactor which liberates 60 times as much energy from uranium as thermal reactors.

The amount of depleted uranium in or discharged from thermal reactors already in the UK could be equivalent to some 20,000 million tons of coal—nearly half the country's estimated recoverable reserves of coal—if used in fast reactors.

The document says that the economic and environmental impacts of *alternative sources of energy* from tapping the energy of the sun and moon—solar, wind, wave and tidal power—need to be fully assessed because they could make a useful contribution to energy supply as conventional energy prices rise.

Energy from *waste materials* has very limited potential; that from *geothermal* power is much greater but there are serious geological and technical problems; and, while *fusion* power cannot contribute to the UK's energy resources over the next 50 years, its energy potential is so great that development of the technology must continue.

Conversion and distribution

The document says that there is little prospect for improving the efficiency with which primary energy is converted to electricity from the current 30–35 per cent to much over 50 per cent.

The conversion of existing power stations to combined heat and power schemes (CHP) is unlikely to give significant immediate benefits. But long term the incorporation of combined heat and power plant in large district heating schemes could be much more attractive.

There is a 'well established case' for installing CHP schemes in industrial establishments with large, steady demand for both heat and electricity. There are thought to be about 300 private CHP schemes already in the UK meeting 20 per cent of industry's demand for electricity, but there is scope for the extension of CHP into the small industrial and large commercial sectors.

Energy use

The efficiency with which energy is used in buildings (mainly for space and water heating) could be improved by the application of existing technology; heating efficiencies could be raised from a present average of 40 per cent to perhaps 60 per cent.

Much of the potential savings could be realised in 15–25 years through changes in regulations and improvements in equipment and specifications. But improving efficiency with which energy is used is only one way of saving energy. It is equally important to reduce the rate of increase in the need for it.

This is particularly effective because on average $2\frac{1}{2}$ units of primary energy are saved in saving one unit of energy consumed at the point of use.

Thermal insulation of existing buildings, better energy housekeeping and management, using less material or materials with a lower energy content and switching modes of transport can reduce the need for energy. Longer term energy savings of 50 per cent or more could be achieved by designing buildings with a lower energy requirement while still allowing for a better lifestyle.

SCENARIOS

The document says: 'The UK's future technology requirements in energy are inextricably linked to future events, most of which cannot be forecast with any degree of certainty; yet one of the main purposes of an energy R&D strategy must be to ensure that future policy makers are equipped with appropriate energy technologies to implement their policies whatever shape the energy economy takes. Therefore, it is clear that we need to adopt an analytical approach which assesses technology needs over the spectrum of possible futures, and not just on some central view.'

Of the many techniques available, the study adopted an approach in which the contribution of each technology is assessed over a set of narrowly defined 'snap-shot' views of the future—termed 'scenarios'. The scenarios are not intended to represent preferred views of future events; they are illustrative and should be seen as a representative set of views which adequately span the range of possible futures and which facilitate discussion of other perhaps less likely futures.

The following summarises the prospects under each scenario:

Trends continued: self-sufficiency maintained until the late 1990s, provided the nuclear programme is geared to 50–60 GW of installed capacity by AD 2000. Beyond 2000, with the coal industry stretched to produce 150 million tons, the UK could be importing fuel at a higher rate than ever before unless installed nuclear capacity exceeded 100–120 GW by AD 2010. Limitations on the supply of uranium would almost certainly require the large scale introduction of fast reactors soon after 2000. The current installed nuclear capacity is about 5 GW.

Low growth: with moderate conservation measures and an expanded coal industry, no more than 75 GW of nuclear plant would need to be installed by AD 2025 if imports were to be held below existing tonnages. The large scale introduction of the fast reactor would not become a necessity until the first decade of the twenty-first century.

Limit on nuclear: even with the most restrictive conservation measures and an expanded coal industry annual economic growth rates as high as 2 per cent could not be sustained; the pace of economic development would fall before the turn of the century to what were previously considered to be unacceptably low levels.

High energy cost: This explores the consequences of a further sharp rise in world energy prices before 1980. Conservation measures which stop short of seriously affecting the country's lifestyle could prolong self-sufficiency well into the twenty-first century at UK growth rates of up to $2\frac{1}{2}$ per cent a year. But at least 40–50 GW of nuclear plant would need to be operating by AD 2000 and double that by 2010. Large scale introduction of the fast reactor would be needed during the first decade of the next century.

Sharp rise in prices in 1990s: the resulting crisis could be much more damaging and prolonged than the recent one. With a growth rate after 1990 of 2 per cent a year over 100 GW of nuclear plant would be required by AD 2010 if imports of fossil energy were to be kept below 100 million tons.

Self-sufficiency: to maintain an annual economic growth rate of 3 per cent over 100 GW of nuclear plant would need to be installed at the turn of the century and this might not be practicable if fast reactors were not operating on a large scale by AD 2000. This scenario makes it essential to press ahead with conservation policies, especially those which reduce fossil fuel consumption, and to expand the coal industry.

High growth: assuming an annual growth rate of 4 per cent, UK self-sufficiency would not persist into the 1990s without vigorous conservation measures. Soon after the turn of the century imports of energy would exceed current tonnages even if 100–110 GW of nuclear plant were installed by AD 2000 and double that by 2010. A nuclear building programme of this size might be possible with UK technology if another tranche of thermal reactors were ordered in the early 1980s and if fast reactors were operating commercially before the mid-1990s.

IMPLICATIONS OF THE SCENARIOS

The document makes the following points:

Primary energy

—considerable attention needs to be paid to coal utilisation and conversion technologies;

—the medium term importance of North Sea oil and gas justifies high priority on the development of Continental Shelf technologies, including enhanced recovery techniques; in the slightly longer term technologies to develop deepwater finds (over 200 metres) may become of similar importance;

—we must be prepared for an expansion of nuclear technology if we are to maintain the option of keeping our energy costs competitive with other countries; the extent to which we take up the nuclear option will depend to a considerable degree on the social acceptability of nuclear power and on judgments about the effectiveness of safety measures;

—any shortfall in fossil and nuclear primary sources of energy would greatly increase the need for alternative sources and this implies a wider R & D programme on primary energy than might be justified by economic evaluations at today's prices.

Conversion and distribution

—some new technologies will be required to produce substitute natural gas and hydrocarbon liquids from coal and for energy storage and advanced generating systems; but in many cases these developments will be available from other countries in advance of UK needs or are being developed by the industries concerned;

—the improvement of the efficiency with which primary energy is converted to electricity will be of continuing importance.

Energy use

—energy saving can contribute as much to the UK energy economy as a new source of primary energy and R&D to achieve such savings deserves high priority; the major opportunities are in buildings, industry and transport and a prime objective must be to develop economically competitive equipment and processes which can be made widely available and attractive to the average user.

The emerging R&D strategy

The document considers the most appropriate course of action for each technology, taking into account the importance of the technology, the possible target dates, the phases of technical development and the availability of resources for R&D.

It outlines research priorities and the main implementation options.

Future work and broader implications

The document states that, generally speaking, more work needs to be carried out on time scales and contributions, on commercial production costs and implications for industrial infrastructures, and on the resulting demands placed on R&D resources by different R&D strategies.

It also draws attention to the non-technological factors which influence the implementation of advances in technology.

Consumer Groups

Energy for industry—the industrial consumer's view (Confederation of British Industry)

Introduction and summary

1 Within its membership the Confederation of British Industry (CBI) covers a wide span of energy producing and supplying industries, both nationalised and private; their suppliers of goods and services; and energy consumers from every sector of industry.

2 For purposes of the Energy Conference it has been agreed by the majority of these interests that the CBI should predominantly represent the views of industrial energy consumers but that these should cover the full range of topics suggested by the Secretary of State for Energy, including energy prospects for demand, production and supply, policy and planning, and relationships between Government and industry.

3 This paper accordingly sets out the industrial consumers' views, fears and hopes for energy prospects to the turn of the century. It suggests some objectives and criteria for a national energy policy designed to satisfy the needs of energy consumers through stable, well-developed and efficient energy industries. It supports the idea of co-ordination and consultation on energy policy and planning through joint discussion between the managements and unions of the nationalised energy industries and Government, but it reaches the conclusion that this tripartite form of consultation alone would be incomplete and could become introspective and self-centred.

4 It therefore recommends that consumer consultation should become an essential part of the formative stages of energy policy formation and energy planning. The tripartite triangle should be squared by the addition of informed and active consumer participation in equal partnership with the three already-proposed participants. Existing organisations for consumer representation, although very necessary and useful in their particular way, are not felt to be suitable for this. The paper suggests the examination of other ways of securing effective consumer participation.

ENERGY PROSPECTS FOR THE NEXT TWENTY-FIVE YEARS

Current development plans

5 The development of the energy supply industries and the exploitation of conventional sources of energy until around 1985 are covered by existing plans and forecasts as reflected in the National Coal Board's *Plan for Coal* and the *Reports of the Coal Industry Examination 1974*; in the Secretary of State's Report to Parliament on the *Development of the Oil and Gas Resources of the United Kingdom 1976;* and in the nuclear power station ordering programme based upon the Secretary of State's Report to Parliament on *Nuclear Reactor Systems for Electricity Generation* 1974 (HMSO).

6 The CBI has already welcomed the plans which have so far been made and the good auguries for development of our resources of coal and off-shore oil and gas. The coal industry is presented with a great opportunity for revival, whilst the offshore operating companies are now moving deservedly towards a hopefully successful outcome of their earlier foresight and skill in locating and developing our undersea resources.

7 In spite of the small and tentative nature of any programme attending the choice of a new reactor, the decision to embark upon a new nuclear power programme after so long a period of delay and disappointment was welcomed by industry as being a very necessary step forward in the direction of greater use of nuclear power for electricity generation so as to conserve fossil fuel resources. The increase in the nuclear component of the electricity fuel mix which could follow the eventual adoption of a larger, long-term nuclear power programme based upon successful and safe initial development of the SGHWR, would also confer a useful added degree of flexibility on the electricity supply system.

Limitations of the current plans

8 The room for extension or alteration of these plans or programmes within the present decade is

limited by factors of available time, finance, manpower, and the engineering and production capacities and capabilities both of the energy industries and of those industries who serve them. However, taken in combination and given the best of progress, these measures are regarded by the Government as being capable of meeting the country's energy demand from indigenous resources in the early 1980s. In the meantime, a finely-balanced and inflexible energy supply situation, in which importation will still figure prominently, could prevail—even when seen against a background of low demand growth and an increasingly effective conservation effort. In these circumstances, any failure on the part of one energy industry to progress steadily towards its target could not be countered by an increase of capacity and production in another—even if the products were wholly interchangeable.

9 There also appears to consumers to be little, as yet, in the country's energy plans (particularly in respect of the nuclear power programme or the preliminary planning of a second phase of development of our coal and offshore oil and gas resources) to provide a springboard from which to expand production to meet any growth in demand during the latter half of the 1980s and even beyond. This second phase of development, for which preliminary plans should surely be set in motion now, is felt to be necessary for providing a bridging period of continued conventional usage while advanced methods for the conservation and better utilisation of fossil fuels and their eventual substitution or supplementation by other energy forms are being developed and brought to fruition.

Planning for the future

10 At the time when the plans were first announced, and in the light of succeeding changes in production forecasts, industrial consumers felt considerable doubts about the wide area of probability which existed, and has continued to exist, between the upper and lower limits of the targets mentioned in the plans.

11 However, since that time, energy demand has become substantially depressed by the intervening economic situation and by the effects of energy conservation measures dictated as much by price as by physical necessity. For the present, therefore, industry's concern about the adequacy of longer-term energy production targets may appear to be unfounded. But considerable uncertainty still attends energy demand forecasting and, if there were to be any substantial increase in energy demand (say from a rapid and sustained upturn in the economy between now and 1980), industrial consumers' earlier doubts about the difficulties of the energy industries attaining the upper limits of production mentioned in the plans could well be realised. These doubts extend to the following features of the plans and forecasts:

(a) The plans for coal would appear to aim at stimulating a demand for coal possibly up to as high as 150 million tons a year in 1985, but the figures given in the *Coal Industry Examination Interim Report* would appear to add up to only 137 million tons (122 deepmined + 15 opencast). The remainder is seen largely as arising from increased rates of production within these redeveloped or new resources. This will obviously depend upon the coal industry's ability to increase and maintain its productivity accordingly.

(b) Correspondingly, in the Government's most recent forecasts for North Sea oil production, the forecast of production for the year 1980 is seen to lie between 95 and 115 million tonnes, whilst that for the 1980s is between 100 and 150 million tonnes a year.

(c) Forecasts for gas production from the North Sea during the same time are stated at around the 6,000 million cubic feet a day (mcfd) mark, but in terms of what the reserves could support rather than what could necessarily be expected.

12 These kinds of cautious approximation are not unusual in forecasting but, from industry's point of view as a consumer, margins in expectations of the order of 13 million tons of coal a year, 20 to 50 million tonnes of oil a year and, say, 1,000 million cubic feet of gas a day are reckonable causes for uncertainty in industry's own future fuel and energy planning.

13 Leaving aside these concerns for the more immediate plans and prospects, there is an awakening concern in industry regarding the prospects for energy supplies in the late 1980s and beyond. This concern extends to:

(a) The ability of the coal industry to continue expanding on the basis of present mining technology, the demands that it makes on manpower, and the potentialities for coal if viewed solely as a conventional fuel. An imaginative programme of research and development is already in train in the coal industry but the backing of further research into methods of coal production, processing and conversion, and use both as fuel and feedstock in the longer-term seems necessary.

(b) The increasing technological difficulties and diminishing economic feasibility, in the light of increasing development costs, of developing offshore oil and gas resources beyond the present 'commercial' fields into areas as yet regarded as being only probable or possible subjects for development. The need to develop favourable circumstances within which the offshore operators can continue to operate for reasonable returns and, incidentally,but importantly, to provide work for the supporting industries, is seen to be essential.

(c) The need, briefly mentioned before, to plan and make advance provisions for second phases of development of our coal and offshore oil and gas resources, at the same time as the current development phases are in train, so as to ensure continuity of supplies and to bridge the period while new energy techniques are brought to bear. The attendant demands on finance, suitably qualified manpower, and scientific and engineering capacities and capabilities need to be anticipated now.

(d) The need, again concurrently, to plan for the eventual conservation of diminishing reserves of fossil fuels through better techniques for their utilisation, and for their substitution initially by nuclear power pending the development of other substitutes or supplements from new energy sources and forms. In this connection, an early review of the current nuclear power programme with a view to its eventual extension would seem to be necessary.

(e) Finally, in extension of (d) above, but no less importantly, the need to provide substantially for a large and wide-ranging research and development effort towards new and 'benign' energy forms. A start has been made on this with the recent grant for research into wave energy, but this and other deserving forms of research will need to be increasingly heavily backed. In this connection, the possibilities for international co-operation and mutual assistance through the strengthening of EEC and IEA relationships should not be decried.

OBJECTIVES AND CRITERIA FOR A NATIONAL ENERGY POLICY

Objectives

14 The CBI has always had some reservations about formal, highly directional kinds of energy policy—possibly from experience of the way in which the turn of events overtook and made nugatory much of the policy followed as the result of the 1965 and 1967 White Papers, and because of exposure to some of the earlier, highly dirigistic, proposals for an EEC energy policy which were put forward by the Commission before and immediately subsequent to our entry into the Community.

15 However, the needs of the current and future energy and economic situations obviously call for a high degree of co-ordination and support of the efforts of the energy-producing industries to encourage and facilitate their development towards greater efficiency and productivity, and to ensure the production of energy supplies in quantities and forms which are adequate and suitable for the country's well-being and economic progress.

16 Industrial consumers would undoubtedly favour this kind of co-ordination and support as a basis for an energy policy. Put very broadly, the objectives of a United Kingdom energy policy as seen from their point of view would be:

(a) To ensure the production or acquisition of adequate supplies of fuel and energy of all kinds to meet the needs of all classes of consumer, with fairness for all and at a supportable cost to all.

(b) To encourage and assist the energy producing, supplying, and distributing industries to meet these requirements effectively and efficiently, and to ensure that the industries and those who work in them receive a fair rate of reward for the materials and services they provide.

(c) To encourage and assist energy consumers of all classes to make the best use of available fuel and energy.

(d) To encourage and assist all forms of exploration, research and development necessary for (b) and (c) above and for the discovery, development and eventual bringing into production of new resources and new and improved techniques of energy production and utilisation.

Criteria

17 The institution and execution of a policy to fulfil these objectives should not entail a high degree of direction and intervention by Government, but it would call for considerable and skilful co-ordination, consultation and monitoring to ensure that the energy industries moved smoothly and harmoniously towards the attainment of their individual production objectives. These objectives, both individually and nationally, would need to be kept under continuous forward review to ensure that they remained realistic in relation to consumers' future needs and to supply capabilities, and that all necessary planning provisions were made to ensure their attainment.

18 Under such a policy the private sector energy industries would certainly expect to exercise

their accustomed degree of managerial and commercial freedom, and there is no reason why the nationalised energy industries should not do likewise subject to the limitations of their special status and public accountability. Within the boundaries of the national objectives, room for healthy competition between the energy industries in obtaining their share of the energy market should exist. Obviously, too, full opportunity should still be offered and taken for exporting where the balance of national advantage as between home consumption or overseas sale was seen to lie that way. Correspondingly, importation should not be excluded as a means of satisfying demands for which the home product was either insufficient, patently uneconomical to provide, or unsuitable.

Industrial consumers' particular views on energy policy criteria

19 The foregoing observations on objectives and criteria for a national energy policy are consistent with the CBI's general views on energy policy principles as set out in CBI *Paper T 19 75* of January 15, 1975 and as earlier expressed in an EEC context in CBI *Paper T 37 73* of February 14, 1973. These papers reflected the views of the CBI's energy supplier members as well as the views of consumer members. It is not proposed to repeat the views at length in this present paper but, paraphrased briefly from both papers, they are broadly as follows:

(a) A policy for developing UK energy resources so as to reduce dependence upon imported energy is to be encouraged. The aim of the policy should nevertheless still be to produce energy at the cheapest possible price consistent with viability for the suppliers and security for consumers.

(b) In the management of such a policy in the national interest, a measure of Government intervention and control may well be necessary but this should not be allowed to inhibit freedom of choice by consumers, fair competition among suppliers, and the reflection of the true costs of production and supply in energy prices to all sectors of consumption.

(c) A policy of realistic and transparent energy prices should not just visit itself upon consumers in the shape of increased charges. It should be directed towards promoting efficiency and competitiveness among energy producers. There is no reason why the energy industries, assured of security and good development prospects within a beneficial energy policy, should not be prepared to compete with each other on equal terms.

20 Complementary to the above, the industrial energy consumers hold the following particular views on energy policy criteria:

(a) *Security of supply:* Security of supply, but not wholly regardless of the accompanying cost, is a paramount consideration. In the consumer's mind this does not just imply freedom from the strategic or economic vulnerability of imported supplies as conferred by indigenous fuel policies, stocking policies or other safeguards. Nor does it just mean the removal of threats of industrial action against either home or imported supplies. Important though these two aspects of security are, the consumer is equally concerned with security in the sense of continuity and adequacy of supplies—hour by hour, week by week, and year by year in normal everyday circumstances.

(b) *Scale of availability:* Equally with his need for secure continuity of supply, the industrial consumer, particularly, needs to feel sure that his chosen energy or fuel will be available on a sufficient scale to meet the expanding needs of his own business. Without this assurance he cannot plan—and it should be remembered that the lead times for his expansion plans will not necessarily be dissimilar from those of the energy industries themselves. He is therefore vitally interested in the development of an energy industry, in the projected increases in its production capacity, and in the location and development of new sources of supply. He therefore needs to feel that timely and adequate plans for expansion of the energy industries and their resources will be taken care of in the national policy. He also needs accurate and continuing forecasts of availabilities to take into account in his own future production planning.

(c) *Freedom of choice:* The industrial consumer plans to stay for a long time with his chosen fuel or energy, not solely out of preference, but because he has provided himself with expensive plant and equipment to suit, and has adapted his processes accordingly—hence his interest in the continuity and increasing availability of supplies as factors of choice. His choice may also extend to the provision of alternative, standby, facilities based upon another fuel or energy supply which he might expect, not unreasonably, to be made available to him at all times on non-discriminatory terms. Furthermore, when the time comes to replace obsolete plant or equipment or to plan a new project altogether, he will reassess his experience with his earlier choice of fuel or energy and may elect for a change. In the exercise of all of these opportunities for choice he could stand to benefit from the kind of energy policy which would ensure that there were no 'Cinderellas' among the energy industries yet would place no undue constraints upon their freedom to go into the market-place and sell at their own competitive prices to free-choosing customers.

(d) *Competitiveness of prices:* As with availabilities, the industrial consumer needs good forward indications of price trends to help him with his own planning. A well co-ordinated and monitored energy policy could possibly help here, from the point of view of bringing a degree of stability and predictability to price trends. However, this should not extend to the introduction of artificialities into prices to produce non-competitive norms. The industrial consumer has subscribed to the CBI's view that energy prices should reflect the true costs of production and, following the justifiable phasing out of Government subsidies to the nationalised fuel and power industries, he and other classes of consumer have experienced some sharp price increases. With understandable consistency, therefore, he does not favour a policy of price distortion and discrimination by the building of taxation into energy prices or the linking of energy prices one to another for the purposes of equalising competition among suppliers. To him, these kinds of price manipulation represent an unwarranted increase in his own industrial costs and a reduction of his own competitive power accordingly.

Environmental considerations

21 It does not come within the compass of this paper to attempt to establish criteria for the environmental aspects of energy policy and planning. Nevertheless the CBI believes that a responsible and constructive attitude towards protection of the environment should be an essential part of policy formation and planning, both for the production of energy supplies and for the use of products either as fuels or feedstocks.

22 In matters concerning the possible impact on 'the environment of industrial products or wastes arising from industrial processes, the CBI would expect to continue its previous and present co-operation with Government on the consideration of appropriate legislation and its means of implementation, and with Government-appointed bodies in the examination of particular aspects and problems. Through these extensive consultative procedures, which are a feature of the Department of the Environment's activities and which involve liaison with the Department of Energy and other interested departments, the United Kingdom's present reasonable and rational approach towards protection of the environment will doubtless be continued.

GOVERNMENT AND INDUSTRY RELATIONSHIPS IN ENERGY POLICY AND PLANNING

The Government's role

23 It has already been suggested in para. 17. of this paper that the Government's role should be one of consultation and co-ordination in its relation to the management of the energy industries, and not of direct intervention in that function. Obviously the extent and influence of such a role would bear more significantly upon the nationalised fuel and power industries and upon the State oil company than upon the private sector of the oil industry. Even in the latter case, the merits of consultation are already well understood between the Government and the oil industry, and recent legislation (however strongly contested and unwelcomed by the oil companies) now provides something of a framework for future co-ordination and co-operation.

24 Over and above this, however, the Government has an essential role to play in bringing about the kind of economic climate within which the fuel and power industries can make their maximum contribution to the strength of the economy and to the easing of the balance of payments situation.

25 Commenting upon the Government's role in an answer to a Parliamentary Question on February 16, 1976, the Secretary of State for Energy referred to the key role of government in developing an integrated energy policy in conjunction with the industries and those that work in them. He stressed that the interests of domestic and industrial consumers, long term as well as short term, should be safeguarded along with the interests of supplying industries and the community as a whole.

26 Industrial energy consumers naturally echo the Secretary of State's wish that their interests should be safeguarded, and the last section of this paper is devoted to the discussion of ways in which this could be better done.

The suggested new machinery for consultation and co-ordination

27 Speaking later, on March 3, 1976, to the National Executive Committee of the Transport and General Workers' Union, the Secretary of State

outlined a three-point plan for long-term energy planning. The elements of this were:

First:

A regular tripartite pattern (management, unions and government) within each fuel industry to monitor its work and discuss its plans.

Second:

A national energy forum including unions and management in the fuel industries, together with others, to go over the published reports issued by each of them and to subject to critical examination the strategies proposed.

Third:

A regular procedure for publishing reports on the Government's progress and plans so that Parliament can exercise a closer scrutiny over what is done and what is planned.

Comments on the suggested machinery

28 The acceptability of the tripartite form of consultation and the degree to which it could become an effective part of the energy industries' own planning arrangements is a matter for the parties concerned to settle among themselves. However, seen from the wider aspect of the management of the country's energy affairs, there is grave concern among industrial consumers that, without corresponding effective and continuous consumer participation or consultation, this form of planning discussion could result in plans and policies which were producer-orientated and in which the consumer considerations became largely what the three parties thought they should be, rather than what the consumers know them to be.

29 The national energy forum, of which the present Conference is presumably the forerunner, would serve a limited purpose as a medium for bringing strategies, plans and performance into the open for very broad consideration by consumers and other interests but it would not provide a means of bringing consumers' practical views to bear in the *formative* stages of policy making and planning. Such a forum would presumably be held infrequently, possibly only annually, and would be too large and disparate to be an effective aid to planning. Furthermore, the consideration of published reports would essentially be 'post-facto' and would be too late for the forum's views to be taken into account other than in the next, distant, phase of planning. Nevertheless, it would be informative for such an open forum to be held. The third feature of the proposals, namely the publishing of reports for Parliamentary consideration is considered by industrial consumers to be wholly desirable in the national interest.

SUGGESTIONS FOR IMPROVED CONSULTATION WITH INDUSTRIAL CONSUMERS

The claims of industrial energy consumers to participate in energy policy making and planning

30 Reference has been made in para. 24 of this paper to the need to enhance the contribution which the fuel and power industries can make to the economy. In this connection, and in relation to national planning, it is all too easy to regard these industries as producers rather than suppliers. Given the opportunities which present themselves to the energy industries for exporting their products directly, their major contribution to the economy must still lie in their role as suppliers to those sectors of industry and commerce who, by nature of their particular products and services, are the direct earners of the nation's wealth.

31 Industrial energy consumers in the private (general industry) sector occupy a predominating position among these. They absorb something like 32 per cent of the country's energy input (the largest single sector of consumption) and they produce around 60 per cent of the country's export earnings. Their claim to a voice in the country's energy planning is therefore undeniable. It is their view that this voice should be heard at the time when policy and plans are made and that a strong need exists for the incorporation of consumer representation into the tripartite mechanisms suggested by the Secretary of State.

32 It remains, therefore, to examine existing forms of consumer consultation, the opinions of industrial energy consumers on them, and to suggest a better form of working participation in which informed consumer opinion, coupled with useful indications of consumers' own plans, could be brought to bear in the formative stages of energy policy making and planning. Such a form of working participation would complement and help the work of the tripartite planners and would provide a means of cross-reference and co-operation in the prospective intervals between the national energy forums.

Existing forms of consumer consultation

33 Existing consumer organisations such as the Electricity Consultative Councils, the National Gas Consumers' Council and the Regional Gas

Consumers' Councils seem to be mostly occupied, possibly justifiably, with end-matters of energy consumption as seen from the viewpoints of individual energy consumers. The CBI's industrial consumers, whilst acknowledging the continuing usefulness of these Councils in individual cases, do not regard them as a suitable medium for the representation of industrial consumers' views on major matters of energy policy and planning.

34 The remaining forms of representation and consultation at present open to industrial energy consumers are:

(a) Individual representation; and representation through the CBI or though organisations such as Trade Associations and Employers' Federations.

(b) National representation, in one energy sector only, through the CBI's Industrial Coal Consumers' Sub-Committee—an organisation set up at the request of the Government, and with the agreement of the National Coal Board, to replace the former statutory Industrial Coal Consumers' Council, whose continuation, under the EEC's rules, was not possible.

(c) The Joint Energy Policy Discussion Group between the CBI and the Department of Energy, at which energy policy matters are discussed from time to time at senior Departmental level.

None of these, as they stand, provides an effective medium for bringing industrial consumers' opinions to bear in the formative stages of energy policy making and planning, although the interests represented are wide-ranging and influential.

Suggestions for the future of the Conference and for improved consumer participation in energy policy making and planning

35 The Secretary of State's suggestion for a national energy forum, in spite of its very real limitations as a potential means of consumer consultation, has already been welcomed in paragraph 29 of this paper. It is possible that a means of continuing consultation with consumer and related interests for purposes of energy policy formulation and planning might be provided by perpetuating the Conference in a concentrated, working, form during the interval between plenary sessions. For example:

(a) The Conference could remain embodied as a consultative organisation.

(b) A small, but highly expert technical secretariat could be appointed.

(c) The purpose of this secretariat would be:

(i) To take the opinions of each successive Conference and to pursue them with the Government and with the managements of the fuel and power industries.

(ii) To provide a continuing means of liaison with the tripartite groups or with the individual managements of the energy supply industries, so as to ensure that consumers' views are brought to bear at the earliest possible stage in national energy planning.

(iii) To establish contact with all consumer and related organisations to ensure that their views and information are fully collated, understood, and presented to the tripartite groups.

(iv) To organise successive Conferences.

36 In this way, the suggested secretariat could provide a continuing and useful input of consumers' views, consumer-information, and other Conference thought to the energy industries' planners and to the Government, either as a matter of course or on demand. It would be available for immediate consultation by the planners, the Government, and the established consumer organisations and other Conference members, thereby providing the fourth essential link in energy policy consultation, co-ordination and advance planning.

37 The extent to which the secretariat covered domestic and industrial consumers' concerns and other Conference interests would be for the Conference to decide—but, seen from the aspect of the needs of industrial consumers alone, a skilled representative organisation of this nature could be a valuable adjunct to the Secretary of State's suggested 'three-point plan'.

38 Alternatively, and seen from the viewpoint and interests of industrial consumers alone, the possibility could be explored, say in conjunction with the Department of Energy and the energy industries, of strengthening and coalescing the several forms of representation mentioned in paragraph 34 of this paper into an effective kind of industrial energy consumers' consultative body.

39 It will be appreciated that the foregoing suggestions are tentative only, and that any part which the CBI might play in examining them further or helping to put them into effect would be contingent upon the prior approval of the Confederation's members. Nevertheless a real need to bring informed consumer opinion from industry to bear at the earliest possible stage in energy

policy making and planning has been demonstrated, and this need should be met.

CONCLUSION

40 The calling together of this Conference is an earnest part of the Government's desire to bring energy consumers and other interests more closely into the framework of energy policy formation and energy planning. Industrial energy consumers, through the CBI, are appreciative of the opportunity to take a part in the Conference. They hope that the spirit of it can now be translated into continuing action aimed at bringing about timely and effective consultation with the consumers when energy policy and plans are under discussion.

Fuel pricing policy (National Consumer Council)

No one who expresses opinions on the policies which should be adopted by the energy industries can fail to have some sympathy for them. Everyone has been hit to a greater or lesser degree by the same calamitous inflation. But the energy industries, and their consumers, have been hit hardest. The rise in oil prices and the elimination of subsidies were not of the industries doing. But we have to do what we can to cope with the problem, and unconventional options which would not have been even considered a few years ago may now be worth serious attention. It is with this introduction that I put forward the case for an option which I fear the industries may not wholly like. I can only hope they will consider it on its merits and not do so in a traditionalist spirit.

This paper looks at fuel pricing policy in the electricity and gas industries as it affects the domestic consumer, and argues the case for uniform flat rate tariffs to replace the existing tariff structure.

Existing tariffs

There are two types of electricity domestic tariff, standard and off-peak. The *standard* tariff is a two part tariff, usually with a quarterly standing charge and a flat rate charge for each unit of electricity consumed. Consumers with prepayment meters pay an extra 50p or so per quarter towards the extra costs of the meters.

There are two versions of off-peak tariffs. The *white meter* tariff is a two part tariff with a higher standing charge (to cover the cost of a meter which charges up different amounts for day or night use) and two unit rates, one for daytime consumption and one (much lower) for night-time consumption. The *restricted hour* tariff is used for appliances wired on a separate circuit which run only at specified off-peak periods, and consists of a small quarterly charge (to cover extra metering costs) and a flat rate unit charge.

There are three domestic gas tariffs. The *prepayment* tariff is a block tariff—the first block of consumption has a high charge per therm and subsequent consumption a lower charge per therm. The *general credit* tariff is a two part tariff with a quarterly standing charge, an initial block of consumption at a high charge per therm, and a lower charge per therm for consumption beyond that. The *gold star* tariff is a two part tariff with a quarterly standing charge and a single charge per therm.

Typical running rates are: 25·7p per therm for the initial block on the prepayment tariff; 19·7p per therm for the initial block on the general credit tariff; and 12·2p per therm on the gold star tariff. Taking standing charges into account, prepayment gas is cheapest up to consumption of 20–25 therms per quarter (about that of a cooker) and the gold star tariff is cheapest for consumptions above about 40–45 therms per quarter (eg a cooker plus limited use of a gas fire).[1]

Consumption

The present structure of electricity and gas tariffs favours large consumers and penalises small consumers. In the case of gas this is quite clearly reflected in the tariff structure. Both between and within tariffs the large user gets his gas at a lower cost per therm than the small user. In the case of electricity, the effect is not so clear from the tariff structure but the existence of a quarterly standing charge means that large users pay a lower average price per unit consumed than small ones. Small electricity or gas users pay nearly twice as much per unit of fuel as large users.

Flat rate tariffs would be 'fairer', because all consumers of gas or electricity—large or small—would pay the same average price per unit of fuel.

Energy conservation

The existing tariff structure is *promotional*—that is, larger users get their fuel at a lower average price than small users.

The need to conserve energy suggests that consumers should not be able to reap any advantages from increasing their consumption. In 1975 a Parliamentary Select Committee[2] recommended that:

'As a minimum the Department of Energy should ensure that electricity and gas tariffs at no point offer lower prices for additional consumption'. Flat rate tariffs fulfil this requirement.

A change to flat rate tariffs might encourage the consumers who found their gas or electricity bill reduced to increase consumption. On the other hand, the increased price per unit of fuel for everyone might deter consumption. The electricity industry apparently takes the view that, on balance, a net decrease in consumption is likely, which is right from the conservationist point of view, and in accord with all the advertising from the industries on the need to save energy.

Flat rate tariffs serve the interests of energy conservation.

INCOME

The penalisation of the small consumer is in itself grounds for concern over the equity of present tariff structures. But the inequity is compounded by the fact that the small consumer is often the low income consumer.

First results from the National Consumer Council's (NCC) own survey[3] show that about 80 per cent of households with incomes of £1,000 or less a year spent less than average on gas and electricity in the last twelve months. Average expenditure for electricity was £90 a year—about £1·70 a week—and for gas £73 a year—about £1·50 a week. Seventy-five per cent of 'poor' households[4] spent less than average on electricity and 68 per cent (of those 'poor' households who use gas) spent less than average on gas. The corresponding figure for households on supplementary benefit was 71 per cent for both electricity and gas.

Figures for each income group are given in the Table below. Percentages refer to households in each income group, eg 50 per cent of households with incomes under £1,000 a year spent less than £40 on electricity last year and 83 per cent spent less than average.

A flat rate tariff would involve the removal of the quarterly standing charge from electricity and gas tariffs. The revenue now derived from the standing charge would be raised by increasing the price of each unit of electricity or therm of gas. This change would reduce the bills of people with below average consumption and increase the bills of people with above average consumption.

The Table gives us a guide to the likely effects of introducing flat rate tariffs. The proportion of households spending less than average would probably be helped by the abolition of standing charges—that is, 64 per cent of all households using electricity and 55 per cent of all households using gas. For low income households the proportion helped by flat rate tariffs would be much greater—83 per cent of electricity consumers with incomes of less than £1,000 a year and 80 per cent of the gas consumers.

So much for the people who would gain. Some, however, would be harmed by the change, and inevitably so since we are not proposing that

Percentage of households in each income group spending less than given amount on electricity or gas

Income group	*Electricity spending in last year less than £40*	*Less than average of £100†*	*Gas spending in last year less than £40*	*Less than average of £75*
Less than £1000	50%	83%	43%	80%
£1000–£1499	26%	76%	33%	68%
£1500–£1999	18%	69%	32%	71%
£2000–£2499	12%	63%	16%	50%
£2500–£2999	7%	62%	24%	58%
£3000–£3999	11%	55%	20%	48%
£4000–£4999	5%*	53%	16%	28%
£5000–£5999	4%*	46%	22%*	36%
£6000 plus	2%*	40%	3%*	9%*
All income groups	16%	64%	25%	55%

	Electricity spending in last year less than	*Less than average of £100†*	*Gas spending in last year less than £40*	*Less than average of £75*
'Poor'**	30%	75%	32%	68%
Supplementary benefit	30%	71%	30%	71%

† nearest grouping to actual average of £90
* based on less than ten households
** households with an income of up to 20 per cent above supplementary benefit rates

the total revenue of the industries should be reduced. The losers would include some people on low incomes—the minority amongst them who are relatively large consumers. For households with incomes of less than £1,000 a year this minority is 17 per cent for electricity and 20 per cent for gas (about one in five). Flat rate tariffs would therefore help the great majority of the poorest households but some would be worse off unless further remedial action was taken.

To see whether the harmful effects of introducing flat rate tariffs could be offset in other ways, we need to be able to identify the low income large consumers. First results from the NCC's survey suggests they can be split into three categories:

(a) *households getting social security benefits:* about one half of 'poor' households who spend more than average on electricity or gas are getting supplementary benefit. It would therefore be possible to offset any harm done through the supplementary benefit system;

(b) *households locked in to one fuel:* many poor households are large consumers because they have to rely on a single fuel for all their space and water heating and cooking. Some of the biggest spenders are in homes which have electric central heating, usually owned by a local authority. The answer to this problem must lie in the installation, wherever possible, of an alternative fuel supply so people have a choice. In addition, households in difficulties should be helped—eg through local authorities—to obtain the more expensive to buy but cheaper to run appliances which could substantially reduce their fuel bills;

(c) *families with children:* other than through the tax or social security system, the only means of helping this vulnerable group is to try to cut their fuel bills. We have already mentioned help that could be given to install cheaper to run appliances. A second form of help might be a country-wide insulation programme to put in draught proofing and basic insulation, thereby reducing the heavy heat losses from many homes. One obvious means of carrying this out would be to grant-aid voluntary groups and local authorities to take part in the Job Creation Programme. A number of such projects are already happening—much more could be done if the government encouraged a country-wide programme which would begin by helping those most in need, including those who use coal and oil rather than gas and electricity.

Flat rate tariffs would benefit the great majority of low income consumers—and many of those harmed might be helped in other ways. Flat rate tariffs would also benefit the majority of consumers generally.

Comprehension

The present tariff structure is not easy for many domestic consumers to understand. This is especially true for gas, where consumers have to elect to go on the tariff that works out best for them. *A uniform flat rate tariff would be very simple to understand, and to apply.*

Some objections

The arguments against flat rate tariffs are that the existing tariffs reflect costs: the standing charge is designed to cover the costs of servicing the consumer irrespective of consumption (eg metering and billing costs) and the price per unit or therm reflects the cost to the industry of their production. Cost-based prices ensure that the consumer's choice between fuels is guided by the costs of supplying those fuels to him and the standing charge discourages people from requesting a fuel supply for standby purposes only. It is more effective to change income distribution through the social security system, taxation policy and so on than through pricing policy. And departure from cost-based pricing leads to arbitrary price-setting and loss of incentives for increased efficiency.

Each of these arguments is looked at in turn below.

Reflection of costs

The ideal allocation of costs to consumers is uncertain. One can argue that costs of providing consumers with a supply should not be allocated to the individual consumer but borne by consumers in general—because the individual supply is an integral part of the fuel supply system. But in any case, there is considerable disagreement about what costs are in practice. The standing charge should—according to the industries—accurately reflect costs of servicing the consumer. Actual charges are inconsistent. For gas, the charge is typically £1 or £1·25 per quarter for consumers on the general credit tariff and £3 or £5·50 for those on the gold star tariff, although the costs involved must be the same for all credit consumers. The equivalent charge for electricity is about £2·20 (standard tariff), although the costs of servicing the consumer cannot be very different from those in the gas industry.

Again, although the cost of producing a therm of gas is the same for all consumers, the price per therm varies according to the tariff the consumer is on, eg from 12·2p to 25·7p in one gas region.

The size of the standing charge clearly influences the unit charge—revenue must be obtained from one or the other. If there is such uncertainty about the desired standing charge, it is difficult to be sure that the unit charge is in any sense the right one.

Government instructions already cause prices to diverge from costs in order, say, to fulfil given financial targets.

It suffices to say that the existing tariff structure does not seem to reflect costs—as identified by the industries—very well. And there is room for doubt over whether costs are being correctly allocated to consumers by way of standing charges and running rates.

Consumer choice

Almost everyone consumes some electricity—because it is the best and most convenient fuel to use for lighting and for domestic appliances. A rise in the price of electricity compared with gas is not therefore going to affect people's decisions to be electricity consumers.

Suppose it is clear to the consumer that gas is much cheaper than electricity for heating. Although it would make sense to install gas heaters, he may not have—or be able to get—a gas supply to his home (this is the case in many rural areas) or he may not be able to afford the high capital costs. He may have no choice at all over fuel installations if he lives in a local authority dwelling. Only 8 per cent of households with a gas supply to the house are not actually gas consumers.[5]

Because everyone uses electricity, the standing charge can be regarded as a 'meter tax' rather than as a device influencing the consumer's choice of fuel. As almost all those who could be gas consumers are, the same argument can be applied to gas. There are practically no consumers left to be discouraged by the standing charge.

It follows that two part or block tariffs are a very imperfect mechanism for influencing consumer choice of fuel.

Income distribution

It is true that direct action on income distribution through taxation and social security policy can be more selective and hence have greater effect than changes in pricing policy. However, there are political and financial limits to the extent of direct redistribution in the foreseeable future. It is unlikely that the government will want to push progressive taxation to its limits, when it wishes to relieve the pressure on skilled workers and middle management. And policies to increase social security benefits substantially or maintain a high level of subsidies conflict with the government's intention to reduce the level of public expenditure. Pricing policy is a practical alternative, or additional, means to influence distribution and need have no public expenditure implications.

A further problem with direct action on income distribution is that the take-up of social security benefits is incomplete—typically between two-thirds and three-quarters of those eligible.[6] Pricing policy, on the other hand, has the advantage of universality.

Efficiency

Spokesmen for the gas and electricity industries argue that, if prices are not based on costs, setting prices becomes arbitrary, incentives to control costs are reduced and yardsticks by which to measure increased efficiency are removed. In addition, prices may mislead consumers as to the true costs of their decisions to consume more or less or any of a particular fuel.

These arguments are not to be dismissed lightly. Consumers have a big stake in the efficiency of the industries, as improvements in efficiency benefit us all. Nor would it be desirable for price signals in the wrong direction to be given to consumers (eg if the aim is to encourage a reduction in fuel consumption generally, we should not be cheapening the price of fuel to everyone).

Flat rate tariffs would not depart so far from costs (if at all) as to distort industry decision-making and forecasting, or give price signals in the wrong direction. It is arguable whether they would involve any greater departure from costs than government instructions to achieve a given surplus or rate of return on assets.

CONCLUSION

Flat rate tariffs would replace the three existing domestic gas tariffs and the general credit tariff for electricity (with extra charges for electricity pre-payment meters abolished). Separate electricity flat rate tariffs would apply to off-peak electricity, spreading over the lower price of each unit the extra costs only of the special metering involved.

Flat rate tariffs are simple to apply and understand. They are 'fairer' in the sense that all consumers of gas and electricity—large or small—pay the same average price per unit. They assist energy conservation. The change would benefit the great majority of low income consumers. And flat rate tariffs do not depart so far from costs (if at all) as to distort industry decision-making and forecasting, or give price signals in the wrong direction. It is arguable whether they involve any greater departure from costs than goverment instructions to achieve a given surplus or rate of return on assets. We hope that their introduction will be given serious consideration.

Notes

1 National Gas Consumers' Council

2 First Report from the Select Committee on Science and Technology, *Energy Conservation.* House of Commons Paper 487, July 1975, p 11

3 The survey was carried out by Social and Community Planning Research and the Centre for Sample Surveys using a stratified random sample of 2,000 addresses throughout Great Britain, preselected from the electoral registers. Interviews were carried out in March 1976 and 1,595 households were seen, a response rate of 80 per cent. Of the 1,595 households, 1,594 were electricity users and 1,106 were gas users.

4 Households with an income of up to 20 per cent above supplementary benefit rates (351 households in NCC survey).

5 NCC survey.

6 Ruth Lister: Take up: the same old story. Forthcoming issue of *Poverty* No. 34, Child Poverty Action Group.

Statement by the Consumers' Association

No one knows what will happen to fuel prices in the long term. We can guess that unless the right decisions are now taken, in our old age the potential demand for energy could so far exceed supply that energy will be severely rationed either by the price mechanism or Government action. Yet for the great majority of people the immediate worry is that they know little about what is going to happen next winter and even less what, if anything, they can do about it. For them it is not a matter of their relationship with the fuel industries, but an almost total lack of communication. The Government urges them to *Save It*. Price is forcing them to do just this but the advice and help they receive is general and often unrelated to their own personal situation and need.

What is the use of telling someone renting a flat with underfloor electric heating that gas would be cheaper? What use telling someone that solid fuel is cheap in their area if their bungalow has been built without chimneys? What use telling someone with electric radiant heaters that gas convectors or electric storage heaters would be cheaper if they have no capital with which to buy and instal the new equipment? This situation cannot be changed by any overnight miracle. We (Government, local authorities, fuel industry, consumer organisations) can see that individuals get advice easily on their own personal problems. We can ensure that in future homes are insulated and built to enable changes to be made with speed from one fuel to another. No house should be built without a flue for the use of solid fuel. People should not be locked into one fuel. We can insist that commercial pressures do not force local authorities and developers to continue to accept favourable terms for giving preference to one fuel or cut capital costs at the expense of running costs. We can seek ways to help the people trapped in all-electric accommodation.

Information for the individual is vital. Hundreds of thousands of homes doubtless waste energy. For many the realistic savings to be made are limited but they should know what they are. The Government and the fuel industries should give the maximum assistance to all forms of local advice centres—Citizens' Advice Bureaux, Consumer Advice Centres, and others. Consumers' Association's own experience has been that the Department of Energy and the fuel industries have been very willing to provide information and help when asked. In fact they have been positively enthusiastic. However, there is a need for much more activity at local level. Could the home economists who work for the gas and electricity industries not bury the commercial hatchet from time to time and hold joint advice sessions in the showrooms? Some people can use general data to make their own calculations. But many others need personal help. This is important. After all, getting heating decisions right can mean the difference between £150 a year heating bills and £500 a year.

There is urgent need for much wider and more open discussions between all the interested people of ways to prevent the waste of energy in homes. The total consumption of energy in homes is higher than in most industries, yet apart from the *'Save It'* campaign more effort appears to have gone into industrial rather than domestic savings. In Germany, a competition along the lines of a 'suggestions box' was launched, to obtain the ideas of builders themselves on how to improve the thermal insulation of existing buildings, and associated costs. Should more work be done in remedying the individual heating installations which lack automatic regulation of the supply of heat? Positive action which could be taken is a raising of the standard of insulation required in new houses and active encouragement of increased insulation in existing dwellings. Again people need personal advice. Most sorts of insulation will save them money. Whether the savings are worthwhile depends a lot on the type of fuel they use for heating, the cost of the particular form of insulation and whether they can afford the capital outlay.

Both the EEC and the OECD have looked at energy labelling of appliances and the Commission's Council of Ministers has before it a draft resolution recommending energy labelling for certain domestic appliances. Work on labels is being undertaken by the appliance industry. Yet all this is happening behind closed doors. There is no discussion of the best way to give people this type of information. There is not even discussion whether minimum standards for energy efficiency would be a better solution in this case. Government and industry should talk to consumers. Further energy labelling of consumer durables must surely come second to telling consumers more about heating appliances. Telling them how much heat they can get out for a given expenditure on fuel.

The industries are in competition. On price this must be right. To have one price for energy, for gas and electricity and coal, would mean distortion of demand, and in the end a higher average price for consumers. However, the idea firmly held by the industries that they must promote increased consumption of their fuel may harm the general interest and the interests of individual consumers. Unless people have the information to enable them to understand first how much energy they will have to buy to use the equipment they are being urged to buy they may make, from their point of view, a foolish purchase. A Code of Practice for the advertising of the fuel and appliance industries is long overdue. What is to be concluded from an advertisement for electric heating with the *'Save It'* stamp and a claim to guarantee running costs?

The conflicting interests of the fuel industries, the certain need for the conservation of existing energy sources, the immediate balance of payments problem and the needs of individual consumers, make policy decisions complex and difficult. A decision which might benefit one interest may have adverse consequences for the rest. Therefore the conflicting claims must be clearly understood and publicly stated and assessed by the Government when announcing decisions.

Government and fuel industries in discussing what fuel shall be used in the twenty-first century must not forget the realities of life in the twentieth. We want facts which will enable us to make more efficient our own use of energy. We want facts to help us join the great debate. We want the consumer councils for the nationalised industries to be given enough muscle for them to be able to apply pressure on Government and industry where and when future policy is decided.

Fuel Poverty (Child Poverty Action Group, Age Concern, and Help the Aged)

Voluntary and statutory organisations concerned with poverty have always been concerned with the problems of the poor in paying for warmth, but the extent of what has come to be called FUEL POVERTY has inevitably worsened with the steep and continuing price rises since 1973. During that winter, when the first important national survey of the heating problems of the elderly was published[1] a number of voluntary groups, including Age Concern, Help the Aged, and the Child Poverty Action Group, came together to form the Heating Action Group. The object of this was to collect information on heating needs of those dependent on supplementary pension or allowance, to provide information and advice about ways in which they might be helped in applying for additional heating allowances, and to argue at all levels that the existing provision for fuel costs was quite inadequate.

Since then, the take-up of heating allowances has enormously increased, adjustments have been made to uprate pensions and other social security benefits in line with escalating fuel prices—although all calculations are still based on an average weighting for all fuels, and take no account of the increasing gap between different fuel costs. There have been some changes, designed to alleviate some of the payment problems of those on very low and social security incomes, by some individual fuel boards and local offices of the Department of Health and Social Security. We are much encouraged by the recent recommendations of the Select Committee's Third Report[2] that too little consideration is given to the user's interests and too much to the convenience of the industries.

But the strain on poor consumers has grown steadily worse, as the experience of our organisations clearly shows in the daily stream of problems over fuel bills that come to us.

Our contribution to this wide-ranging debate on energy policy must therefore be to continue to stress the urgent need for action on fuel poverty. We are aware of the problems of the industries and of the statutory obligation not to discriminate between classes of consumers and individuals in tariff and standing charges; (section 25 (5) of the Gas Act 1972; section 37 (8) of the Electricity Act 1947); we accept the argument that a return to very large-scale flat-rate subsidies is not viable, and indeed that the chief beneficiaries of such subsidies would not, in most cases, be the poorer consumers who are our concern but the big users of energy. But we are here to give some idea of the hardship caused to those groups worst hit by inflation because they must spend all their income on necessities, and fuel expenditure is both unavoidable and rising at a rate far above that of food, rents, or other essentials to daily living.

We shall begin by explaining what we mean by fuel poverty, and what it means for the old, the sick and disabled, the unemployed and families dependent on social security incomes. We shall briefly outline the existing provision made for fuel in those incomes and where, in our view, it fails to meet needs. We then refer to the role of the fuel boards, and end by offering some suggestions as to ways in which we believe fuel poverty might be alleviated.

The events leading to recent steep price rises in all fuels, but especially in electricity, hardly need repeating here. But it may be useful to recall that many of the poorest groups in society, especially the elderly, are only now facing the consequences of our national failure to plan for an era of high-cost energy. The poor elderly still tend to live in homes that were designed without thought of insulation or the capacity for space heating to Parker Morris standards; 78 per cent of the elderly, compared with the national average of 66 per cent, do not have central heating; their homes are often in a worse than average state of repair. Because they have lower rents, poor families are often housed in the oldest accommodation with facilities only for high-cost electric fires, which, along with paraffin heaters and open grates have for too long been the only heating equipment of the poor—largely because they were cheap to purchase.

The first aspect of fuel poverty, then, is inadequacy of heating appliances, lack of insulation, and poor housing using high-cost fuels, which means that very many old people and poor families lack adequate warmth in winter and may suffer hypothermia or its concomitant illnesses and accidents; surveys such as we have listed at the end of this paper, along with many other studies, all show that the chief reason for cold homes is the fear of being unable to afford the fuel bill.

But a new aspect of fuel poverty has arisen—again as a result of the failure of planners to take account of what keeping warm will cost these on pensions and very low incomes. Large estates were designed wihtout insulation, with complicated and highly expensive central-heating systems dependent on electricity.[3] Increasingly, pensioners and poor families have been rehoused without choice of fuel and then faced with very high bills, often without any way of telling in advance how much they would have to pay. And for the poor, unlike the better-off who can move house, or invest in cheaper or more efficient appliances, they have a stark choice—paying up, even if this means going without other essentials, or going without warmth. Fuel poverty, in this sense, means that those who may have managed to budget without a prepayment meter either run into arrears or, by avoiding debt, are deprived of the means to buy food, clothing, or rent.

The groups with whom we are concerned often have higher than average bills, which makes the problem of negotiating payment harder for social and voluntary workers trying to help them, often under threat of imminent disconnection. The reason is obvious: those who are housebound, whether by age, sickness, or young children, or because they are unable to find work, depend on warmth for more hours of more weeks in the year. Hot water consumption is much higher where there are babies, sickness, and limited laundry facilities. All poverty means less freedom to participate in the normal activities of the community, most of which cost money, and warmth in the home may be crucial to morale as well as to physical health.

In August 1975, there were 2,875,000 people dependent on supplementary benefits to bring their pension or other income up to the official poverty level. It is assumed that within the scale rate of supplementary benefit, £2·50 per week is attributable to normal fuel expenditure, including lighting and cooking. No distinction is made for the type of fuel or appliance used, but weekly additions, ranging from 55p to £1·65, may be made at the discretion of the local DHSS, in 'exceptional circumstances'. These may be chronic illness or lack of mobility due to old age, or poor housing that is exceptionally hard to heat. Since November 1975 the powers governing these discretionary payments have been extended to allow *alternative* payments, ranging upwards from 25p to a maximum of £1·10 per week, according to the number of rooms, for those with central heating. Again, the type of fuel is ignored, and a pensioner moving to more limited accommodation but with costlier heating system may thus receive a reduced heating addition. Nor is the presence of young children, severely restricting the parent's mobility, treated as grounds for extra payment towards fuel—the notional £2·50 is within the householder-claimant's basic weekly benefit.

For those on supplementary allowance or pension who pay for landlord-controlled central heating as part of rent, additions are payable to top up any fixed charge higher than £2·50, and it is some indication of the high cost of all-electric council accommodation that in, for instance, the Hulme Estate, Manchester, these additions can amount to more than £3·00 weekly.

The Commission also has power to give lump-sum payments, in exceptional circumstances, to meet urgent needs of claimants. These may include provision for major clothing or furnishing items regarded by the visiting officer as essential, the replacement of outworn, broken, 'totally inadequate' or 'excessively wasteful' heating appliances, or draughtproofing materials—but not anything considered to be 'house improvement'. The Commission is usually reluctant to use their powers to meet fuel bills, regarding this as double provision for those who have failed to save their £2·50.

The number of weekly and lump-sum payments to supplementary benefit claimants has steadily grown with the number eligible because they are below the poverty line. And yet, because of the still widespread ignorance of entitlement, and the reluctance, especially among the elderly, to submit to means tests, it is known that very large numbers do not claim. The inadequacy of the retirement pension and of national insurance provision for certain groups, particularly the disabled and those, such as one-parent families, who cannot satisfy the contribution conditions, means that the 'safety net' of means-tested assistance envisaged by Beveridge has encompassed ever-increasing numbers. And yet still as many as three-quarters of a million pensioners, on a government estimate, may be entitled to but do not claim, supplementary pensions.

The cumbersome machinery of piecemeal extra payments nevertheless seems ineffective in preventing fuel poverty. Those who pay their bills are often left with so little that they must then apply for grants towards clothing or other essentials, and depend on the discretionary decision of one local office, or an appeal tribunal.

Our recent experience of the new arrangements for avoiding disconnection of supply to those on supplementary benefits highlights this aspect of fuel poverty. Agreement was reached with the fuel boards that those claimants defined as likely to suffer hardship without supply may now agree to have direct payments made, deducted weekly from their benefit, to cover current fuel consumption plus an agreed sum towards arrears. The level of these deductions tends to be so high that the claimant is left with insufficient resources for food; one electricity board office, covering about 100,000 households, is currently handling about 2,000 such payments, one-tenth of which are for £10 or more per week. In such cases, a householder with one child may, for instance, be left with about £7.00 for all other living needs except rent.

We have not so far mentioned those who are ineligible for any help under existing social security office policy instructions because they are in full-time work, or have perhaps accumulated a big fuel bill while on a low wage, but are newly dependent on benefits because of redundancy. Referral is often made of such cases to the local authority social services department, who have powers under Section one of the Children and Young Persons Act 1963 to give emergency payments to avoid taking children into care. We do not see the role of the social services as providing income maintenance, and local authority expenditure cuts have recently made such payments far less likely. Nor does settlement of one debt normally solve the ongoing problem of high fuel costs, and there is a high cost in social work time and inevitable delay in application. In the current economic climate, rising demands for fuel bill assistance from charities have become very common, but have less chance of success and clearly provide no long-term solution.

In briefly outlining the special hardship of the poor at a time of rising fuel prices, we have so far hardly mentioned the role of the industries. We know that some local offices and managers have recently been doing their best to reconcile commercial and social considerations, and some positive efforts have been made to deal sympathetically with households where there is obvious hardship. We welcome the recent Government moves to look at tariffs and payment procedures, and have submitted evidence to both committees.

But we anticipate no real improvement in the fuel poverty problem until and unless this Conference leads to some important changes, both in attitude and in legislation. It appears from our experience that unification of the gas industry has considerably reduced the anomalies between individual boards' policies and especially on disconnections, and we would like to see the Plowden recommendations adopted for electricity. We regard the accountability of both industries to the users, at present contained in the consultative and consumer bodies, as woefully inadequate; very few of the poor consumers who most need a voice are aware of how to make use of these bodies, and their powers are either under-used or ineffective in confrontation with the boards.

There are some fundamental points on which action is needed. One reason for the anxiety among poor people over fuel bills is the enforced trend away from prepayment to credit metering—an administrative rather than popular convenience. For those who budget weekly and for whom any debt presents a major problem because they are never able to save, big quarterly bills impose a further burden. Although it is now accepted that rent arrears should not automatically result in a tenant losing his home, fuel supply may be disconnected by administrative action without any judicial safeguard, and for those wholly dependent on electricity there is no realistic alternative. All social workers know of households where the only means of lighting is one calor gas cylinder, candles, or electric torches—all involving much higher running costs—because the bill is too large to be paid and there has been a refusal by the board to negotiate on repayment of the debt by instalment. We welcome the move made between the trade union whose members are responsible for disconnections and the British Gas Corporation to ensure that a house-call must take place to discuss ways of repayment of arrears before cut-off can be implemented. It is in our view essential that all consumers be offered a much wider variety of payment methods, including the option of large-coin prepayment meters—if necessary coupled with small insurance cover against break-in—unless there are overriding safety arguments against this preference. In the short-term, the industries should agree to allow for a period of adjustment to high prices by considerably extending the time allowed for bills and arrears to be cleared—repayment could extend, as in the new agreement for supplementary benefits cases, over a period of up to two years, and a prepayment meter installed and calibrated to cope with current and arrears payments. We would, however, like to see the powers of disconnection for non-payment and the right to demand large deposits as security by late payers severely curtailed by statute. Procedures for fuel debt recovery should be brought into line with those for other debts, through the county courts, at which the debtor has a chance to state his case and agree terms; only after a breach of any such agreement and a further court application by the fuel board should there by any discussion of the possibility of disconnection of supply.

The industries have already begun, but could do far more, to offer effective education in fuel use, advice on heating systems and appliances, and above all to make 'pay-as-you-burn' schemes better known and used by everyone who chooses. In this area, we regard it as essential that the existing role of debt-collecting officer in the board office should be restructured; all customers should have access to a liaison officer who will help to negotiate on current and any outstanding payments and the job of cash collection could, in our view, be shared with outside agents, such as local authority rent officers. Other local or health authority employees who already visit households where there might be payment difficulties could also work with the boards, extending the stamp savings outlets in all areas to local clinics, day centres, and rent offices.

No shift of attitude or policy within the fuel industries can of itself abolish fuel poverty, and while we accept that the industries cannot take on what has been called 'a welfare function', it is unacceptable that they should be operated solely on a competitive commercial basis and shift all responsibility for users' problems or to overburdened social work departments and other welfare agencies. The existing income maintenance provided by supplementary benefits does not seem to have solved the problem, and any discussion of energy costs must be based on consideration of the cost to society in terms of health and social services necessary when fuel poverty worsens, quite apart from the extreme hardship caused to the poor themselves, and conservation.

We accept the general conclusion of the joint departmental report *Energy Tariffs and the Poor** that it is perhaps impossible to devise arrangements whereby selected poverty groups are given fuel vouchers or further special allowances—it is simpler to raise the level of pensions and thus of all social security payments. We regard the extension of the national insurance scheme and a reform of the tax structure to reduce the numbers dependent on means-tested supplementary benefit as the long-term solution to poverty, of which fuel poverty is simply one (not necessarily apparent) feature[4].

In the short-term, we would propose a reassessment of the fuel element in the scale rate of supplementary benefit, based firstly on a realistic weighting for the fuel element, allowing for dependent's needs. The price for pensioner-couples, which is used to calculate upratings, gives a weighting of 13·1 per cent to fuel. This is an average for all domestic fuels, and any family depending on electricity alone would need to spend a larger proportion of their budget. Even if this figure is taken as reasonable, a couple with four children under 10 would be expected to spend £4.12 per week out of a total income of £31.45; the existing £2.50 attributable for all households on benefit will only allow for one 2kW electric fire operating eight hours a day[5]. It is in our view essential that even if a realistic weighting were introduced, there would still be a considerable number who, because of exceptional circumstances, could not afford to pay; in these cases appliances and type of fuel should be taken into account in calculating additional heating allowance.

The role of the local authorities should not be for income maintenance, but we feel they too can help, especially in the long-term battle against fuel poverty. Some authorities make provision for elderly people in kind, by providing insulation, electric blankets etc, under their powers in the Chronically Sick and Disabled Persons Act 1970. We have seen the lamentable record of housing departments in the installation of high cost heating systems, complicated controls and failure to insulate. We welcome the pilot scheme now planned by the Department of the Environment to test economical and effective solutions to heating requirements and costs for the elderly in three Housing Action Areas, and hope that a definite responsibility will be imposed on all planners to balance yardstick building costs against running costs for heating installations.

CONCLUSION

We have seen the consequences for the poor of bad energy planning, inadequate central government advice on heating systems, leading to estates designed for high consumption of high-cost fuels. We must now look to a strategy which ensures that those who have no voice in planning the housing and heating they are offered are not expected to continue to foot the bill for mistakes. While we accept that the cost of subsidising fuel, whether by tariff cross-subsidies or central

*Available from the Library, Department of Energy.

government funding out of taxation, is not under consideration, we regret the negative findings in *Energy Tariffs and the Poor* that there is no case for reforming either the inequitable standing charges or a tariff system which favours large users and penalises the small or medium user on prepayment meter. We would like to see the widest possible choice of payment methods open to all, and major reform of the powers of the boards over deposits and disconnections.

For those with whom we have been especially concerned in this paper, who depend on supplementary allowance and pension, the piecemeal efforts have been too often no solution, and we must continue to press for the extension of National Insurance provision as of right to a level which will reduce the numbers dependent on means-tests and the arbitrary administration of discretionary payments to those who are able or persistent enough to find their way through the maze of the system.

Notes

1 Body Temperatures in the Elderly. *British Medical Journal.* January 27, 1973
2 Para. 184. *Third Report of the Select Committee on Nationalised Industries, Session 1975-76* (HMSO 20p)
3 *Hansard.* House of Lords. March 16, 1976. PQ 599
4 *Social Security: The Case for Reform* by Ruth Lister. CPAG. 1976
5 'Help Yourself to Warmth'—heating bar chart (CPAG and Age Concern) December 1975

Publications and papers on fuel problems

Turning the Heat On, a report of a survey of 83 households by CPAG, Age Concern, Help the Aged and others (June 1974)
Cold Comfort, a survey of heating resources and requirements in 18 low-income London homes (CPAG in association with the Research Institute for Consumer Affairs. July 1974)
Evidence submitted to the Select Committee on Nationalised Industries (Sub-Committee B) May 1975 (HMSO) CPAG, Age Concern, Help the Aged
Evidence submitted to the Lovell-Davis Enquiry (Age Concern) April 1976
Evidence submitted to the National Consumer Council (March 1976) Age Concern
Policy statement on Fuel Debts (February 1976) Age Concern England
Fuel Debts and the Poor (poverty pamphlet 24) by Marigold Johnson and Mark Rowland (April 1976) CPAG
Death in Winter, Help the Aged 1975
Help Yourself to Warmth (December 1975) CPAG and others
Warmth and the Elderly (January 1975) by Dr Ronald Fox (Age Concern)

Towards a national energy strategy (Friends of the Earth and the Conservation Society)

Introduction

The Secretary of State for Energy, the Rt Hon Tony Benn MP, has convened a National Energy Conference, to assist in the formulation of a national energy strategy for Britain. The Conference is a significant step forward. It brings together for the first time face to face the many disparate viewpoints which must be accompanied if we are to reach any common ground on our supply and use of energy. Friends of the Earth and the Conservation Society are grateful for the opportunity to participate.

Objective

Before formulating a strategy it is necessary to have a recognisable objective. We propose an objective for a national energy strategy for Britain. It should aim:

1 to guarantee to every member of the community a supply of energy adequate in quantity and quality to sustain physical well-being and comfort and provide appropriate mobility; and

2 achieve the optimum blend of human skills and inanimate energy in the agricultural and manufacturing industries which fulfil the material and social requirements of the community.

Policy guidelines

A strategy aimed at this objective will involve decisions by central and local government, industry and commerce, trades unions, interest groups and individuals throughout the community. But aggregated decisions do not of themselves constitute a strategy or a policy. They do so only if they are made within a common framework. To this end we should like to propose certain guidelines, against which to measure the energy decisions we face now and henceforth.

1 It is essential to start by considering demand, not supply.

2 Policy can influence demand as well as supply.

3 It is important to know how energy is to be used.

4 It is equally important to know not only the *quantity* of energy to be used but also the *quality.*

5 A distinction should always be made between uses requiring (a) high temperature energy, (b) low temperature energy, and (c) portable fuel.

6 When a mismatch occurs between use and supply it is generally faster and cheaper to reduce demand than to increase supply.

7 Once an end-use has been identified, the thermal performance of the end-use should be evaluated, with a view to possible improvement.

8 Improved thermal performance of an end-use system is usually once for all, whereas added supply at present requires continual sustenance.

9 Ideally, primary energy should be utilised directly at the point or use. However, since this is in practice often difficult, close attention should be paid to the chains of energy conversion from primary to end-use.

10 In choosing between alternative conversion chains from primary to end-use, account must be taken of the amount of primary energy required to deliver a given quantity of suitable end-use energy. Other details being comparable, preference should be given to the conversion chain which requires least primary energy to satisfy an end-use.

11 Resource distribution and conversion chains imply distribution systems; but it should not be assumed that large-scale centralised systems are inherently preferable to smaller-scale decentralised systems. On the contrary, for reasons of flexibility in planning, construction and operation, and overall stability, decentralised systems may now be preferable.

12 Alleged 'economies of scale' must be viewed with scepticism. If demand cannot be accurately foreseen five years hence, and if it takes

ten years to build an 'economically sized' system, the planning process must take place in a vacuum.

13 Whatever the decisions, however they are decided, the future will be different. There is no reason to believe that an electronuclear future will be either familiar or cosy—particularly if it centres on the fast breeder reactor, and involves plutonium as an article of commerce.

14 Forecasts influence policy—but it must not be forgotten that policy also influences forecasts.

15 To express an expected increase as a 'percentage rate of growth' implies an exponential model. Present trends in energy use offer little justification for using such a model. The present economic system requires the majority of end-use energy in the form of low temperature heat.

16 Energy tariffs should not encourage extravagance, especially in industry.

17 When petroleum and natural gas become scarce, the major problem will be that of supplying portable fuels.

18 Electricity cannot readily serve as a substitute for portable fuels.

19 Electricity presently represents only some 10 per cent of end-use energy, and nuclear electricity less than four per cent. Electrical and nuclear Interests should not be allowed to dominate energy policy.

20 Past mistakes—particularly in forecasting, research and development, and investment—impose constraints on present decisions; but they should not be allowed to define future options.

21 It might be desirable to 'keep all our options open'; but we have neither the time nor the resources to do so indefinitely. To support option A may starve options B, C and D. In addition, there may be little justification for supporting all options. For example, the development of *both* coal and nuclear technologies does not appear justifiable in terms of our energy requirements—it may have to one or the other.

22 The proportional support for research and development of different options indicates the anticipated proportions of these options in the long-term energy mix. It is risky to place overwhelming emphasis on a single difficult option.

23 Technologies based on coal and equilibrium energy sources are more likely to generate export business for the UK than is nuclear technology.

24 A production system which is energy-intensive relies less on human skills, offers less employment and is likely to produce what it *can* produce, rather than what the community requires.

The United States Energy Research and Development Adminstration (ERDA) recently published an updated version of its national plan for energy research, development and demonstration—ERDA 76–1—which states that 'Conservation (energy efficiency) technologies are singled out for increased attention and are now ranked with several supply technologies as being of the highest priority for national action.' We strongly endorse ERDA's views on the importance of conservation technologies and believe that some of the characteristics of such technologies singled out by ERDA should be borne in mind in the formulation of Britain's energy strategy. These are:

—'A barrel of oil saved can result in reduced imports.

—It typically costs less to save a barrel of oil than to produce one through the development of new technology.

—Energy conservation generally has a beneficial effect on the environment in comparison to energy produced and used.

—Capital requirements to increase energy use efficiency are generally lower than capital needs to produce an equivalent amount of energy from new sources since most new supply technologies are highly capital intensive.

—Conservation technologies can generally be implemented at a faster rate and with less government involvement in the near term than can new supply technologies.

—Energy efficiency actions can reduce the pressure for accelerated introduction of new supply technologies. Since the actions persist over time, the benefits are continuing.'

Policy recommendations

Partly because of past mistakes and partly because of recent good fortune, Britain in 1976 has an embarrassing excess of energy supply, and a wide range of policy options. We need do nothing hasty. We have time to pause, and to plan with care the direction of a national energy strategy. In this context Friends of the Earth and the Conservation Society would like to put forward the following recommendations.

1 The Department of Energy should publish as soon as possible a detailed description of its own views on energy policy—objectives, options and priorities. It should then be prepared to consider public response to its policy document.

2 The Department of Energy should publish details of its forecasting procedures and results, indicating not only the ranges of possibilities but also those which the Department considers plausible. The forecasts should distinguish between end-uses requiring high temperature energy, low temperature energy and portable fuel. The Department should also publish its model of the energy use and supply system of the country, for general use and evaluation.

3 When any Department other than the Department of Energy publishes either a consultative document on a firm policy statement, they should include an explanation of how the proposals fit in with energy strategy as a whole. The recent consultative document on transport policy was singly lacking in this respect.

4 The Treasury's 10 per cent test discount rate is inappropriate in the context of many promised energy developments, and in any case applied capriciously. Neither the particular rate of 10 per cent nor the overall procedure should be allowed to dominate evaluation of alternatives.

5 The Treasury should revise the yardsticks by which it evaluates public investment, for instance in local authority housing. At present the stringent control over initial capital investment tends to lead to unnecessarily high running expenses. Minimal standards of thermal performance—for instance insulation, necessitated by unrealistically tight yardsticks for construction, lead to excessive fuel costs, which are often borne by occupants who can ill afford them.

6 Upgrading domestic insulation is the most cost-effective form of energy technology now available in Britain. The Government's present position is that such investment is clearly advantageous and should not be given government support. However, many of those in underinsulated homes cannot afford even the moderate cost of added insulation. The Government should: (a) upgrade the Building Regulations such that all new and modernised housing has at least four inches of roof insulation; (b) simplify the Home Improvement Grant procedure such that pensioners and disabled people can get roof insulation without having to wait nine months or more; (c) relax Department of Health and Social Security (DHSS) 'Exceptional Needs Payments' so that those who cannot afford rising fuel bills can at least reduce them by insulating their roofs; (d) advertise (b) and (c) so that people *know* that these grants exist.

7 The Department of Energy should publish as soon as possible its response to the *First Report from The Select Committee on Science and Technology: 'Energy Conservation'*. The Department of Energy's response should give careful consideration to the 42 specific recommendations embodied in the Report. The Department of Energy should either accept and initiate action on each recommendation, or give a detailed explanation of its refusal to do so.

8 Preliminary indications are that the report on publicly funded energy research and development, expected to appear at or before the National Energy Conference, will endorse the present policy of allocating the great majority of funds to nuclear fission, in particular the fast breeder reactor. The Government should reject any such proposal. Research and development allocations tend to be self-perpetuating. It is time to make an explicit redistribution of the available resources, towards a fairer balance. There should be increased support for conservation technology; for coal extraction technology; for the technologies of coal utilisation, especially those based on fluidised combustion; for the equilibrium energy technologies, especially solar energy; and for the technologies of energy storage.

In all these areas it should be emphasised that industrial developments will be of interest not only for indigenous applications but also for export. The export market for the technologies of conservation, coal and equilibrium energy sources is large and expanding. Decentralised systems will be of great value to Third World countries, for which nuclear electricity is in almost every respect utterly unsuited.

9 Since 1954 public funding for nuclear energy has been provided by a seperate Vote of Parliament. This Vote was until recently called the Atomic Energy Vote; but it has of late been redesignated 'Industrial Innovation: Nuclear Energy'. There seems no *prima facie* reason why civil nuclear energy should be thus singled out for separate support, while all other forms of energy research and development have to divide up the left-overs. Consideration should be given to changing the statutory basis of funding for nuclear research and development.

10 In its evidence to the Royal Commission on Environmental Pollution* the Department of

*Royal Commission on Environmental Pollution: Study of Radiological Safety: Evidence by the Department of Energy (1975), available from the Library, Department of Energy.

Energy declared that 'Successive governments have confirmed the policy of developing fission reactors as our main long-term energy source'. Such *ex cathedra* pronouncements simply will not do. The Department of Energy should publish a consultative document describing in detail its view of the role to be played in UK energy policy by nuclear fission, and particularly the fast breeder reactor.

11 The Department should publish a consultative document on the management of radioactive waste from the military and civil nuclear energy programme, past and future: the kind and quantity of waste, and the procedures to deal with it.

12 Despite misleading comment to the contrary, no contract has yet been agreed between British Nuclear Fuels Ltd (BNFL) and the consortium of Japanese utilities, regarding reprocessing of 4,000 tonnes of Japanese oxide fuel. Many important details of this proposal have been inadequately ventilated. Senior BNFL spokesmen have recently contradicted each other, even as to whether the company plans to build one or two new reprocessing plants. Until more is known about the plan, the Government should withdraw its blanket approval for such undertakings and insist on much more information as to technical feasibility, pollution control measures, economics and implications for proliferation of nuclear weapons capability.

13 Industries dependent on orders for power stations now face several years in which no station need be ordered. Rather than devoting an estimated £1,500 million to two prototype nuclear stations, the Government should press for the construction of a series of small coal-fired prototype total energy stations demonstrating various modes of fluidised combustion, on the urban sites of existing stations now earmarked for shutdown.

CONCLUSION

The guidelines and recommendations outlined above are neither comprehensive nor definitive. But Friends of the Earth and the Conservation Society consider them a useful framework for discussion and development of a rational energy policy. We shall welcome comment. We hope that the first National Energy Conference leads to a wider public awareness of the issues, and to a constructive and continuing dialogue.

References

1 *World Energy Strategies: Facts, Issues and Options,* Amory Lovins, Friends of the Earth 1975

2 *Non-Nuclear Futures: The Case for an Ethical Energy Strategy,* Amory Lovins and John Price, Friends of the Earth 1975

3 *Energy Alternatives and Energy Policy: A UK Viewpoint,* Walter C. Patterson and Czech Conroy, Memorandum from Friends of the Earth to the Select Committee on Science and Technology, May 1975

4 *Report of the Select Committee on Science and Technology on new energy sources,* to be published late 1976

5 *Energy Conservation, First Report from the Select Committee on Science and Technology Session 1974–75,* September 1975

6 *Energy Conservation: A study of energy consumption in buildings and possible means of saving energy in housing,* 56/75, Building Research Establishment

7 *Proceedings of the International CIB Symposium on energy conservation in the built environment,* 6–8 April 1976

8 *Fuel's Paradise: Energy Options for Britain,* Peter Chapman, Penguin 1975

9 *Nuclear Power,* Walter C. Patterson, Penguin 1976

10 *The Energy Question,* Gerald Foley, Penguin 1976

11 *Conservation of Energy, Alternative Energy Sources and their Implications for Environmental Conservation and Future Ways of Life,* Gerald Leach, International Institute for Environment and Development, September 1975

12 A Theory of Dimensions, W. L. Wilson, *Energy World* April 1976

Policy

Transcript of the coal and electricity industries tripartite consultations held at Church House, Westminster, on Friday, February 20, 1976

This paper will be published separately as Energy Paper Number 14 by HMSO.

The case for a planned energy policy (Frances Morrell and Francis Cripps)

This paper, which sets out the views of the Political Advisers, was written by them for the Secretary of State for Energy who decided to make it available as a paper for discussion as part of the public debate on energy policy.

Since the 1950s Britain has had a consistent energy policy, based on a free market between fuels, which left users exposed to the risk of unexpected changes in prices and made little provision for long-term security.

This paper examines the implications of this existing energy policy and proposes a more planned, long-term approach for the future.

SUMMARY

1 The traditional policy of free competition on a commercial basis between alternative fuels is reflected in the statutory and administrative framework of the energy sector. The only major exception has been nuclear power, developed by Government decision without applying strict commercial criteria as a safeguard against scarcity of cheap oil and coal.

2 The fundamental question now is whether this policy ensures long-term security and gives confidence that consumers will be able to satisfy their energy needs at a cost they can afford.

3 If energy policy continues on traditional lines, cheap gas will expand its domestic and industrial market at the expense of electricity, jeopardising the coal industry whose main outlet is now in fuel for power stations. Thus the steady build-up of new pits, vital for the long term, will be held back for short-term reasons. Conservation and development of renewable sources will proceed slowly. Offshore oil and gas reserves will be depleted rapidly.

4 Electricity has become very expensive and is likely to remain so because we no longer have a cheap fuel suitable for burning in power stations. Gas is currently cheap in Britain but should be confined to efficient domestic use. It would be folly to waste our limited reserves of this premium fuel in power stations where two-thirds of the heat is lost.

5 High-cost electricity and the long-term scarcity of cheap gas pose a major problem for energy consumers in a country where many families are still very poor and where there is a very large number of badly insulated homes, many of which are all-electric. Many consumers cannot switch to the only remaining cheap fuel—gas—although others will.

6 The projected world scarcity of cheap fuels in a few decades makes it prudent to conserve our exhaustible reserves of natural gas and oil. It would be far too risky to rely heavily on the nuclear solution, when uranium sources, like oil and gas, are scarce and the fast breeder is unproven. To keep our options open we must steadily invest in new coal pits and in coal-fired power stations to make sure the coal can be used.

7 The best long-term guarantee of adequate warmth in the home for consumers would be a combination of parity pricing of electricity and gas, together with a major programme to raise standards of insulation and appliances, both in new and existing houses.

8 Planned development of fuel supplies is only possible if the pattern of fuel use is also planned so that demand and supply for each fuel are maintained in balance.

9 To implement a planned fuel policy the Government would need to determine fuel use in power stations and to fix relative prices of fuels so as to guide the independent choices of final consumers.

10 If the Government is responsible for deciding and implementing a planned policy, the tradition of secrecy in discussion and decision-making must be ended. The Government must present options and consult all the interests involved openly and in public before reaching decisions. It must also make regular reports on energy policy and forecasts.

Traditional energy policy

Although there has been no White Paper on Energy since 1967, a clear policy has been followed with little change since the 1950s right up to the present. The basic assumption of national energy policy has been that the needs of energy users could be met with maximum efficiency by the operation of a free market between competing fuels. This meant that the pattern of energy supply and use in Britain was mainly dictated by the aggregate of choices by individual users rather than being planned by the Government.

The statutory and administrative framework has reflected the basic assumption. Coal, gas and electricity, although organised as nationalised monopolies, were given the duty of supplying in accordance with consumer demand at prices which covered their costs of production. Each of these industries had to compete commercially with the other and with oil imported at the world-market price (subject to excise tax).

The Government's main function was to supervise the prices and investment of nationalised industries to make sure that financial targets would be met and that supply would be expanded in line with the projected growth of demand.

The main exception to the free market principle has been nuclear power. Since the Suez crisis the Government has promoted development of nuclear power without applying normal commercial criteria in order to strengthen long-term security of energy supply. The commitment to nuclear power was justified by the knowledge that reserves of cheap oil and coal are limited and ultimately exhaustible. The free market principle could be over-ridden in the case of nuclear power because of the double coincidence that choice of power stations is under the Government's direct control and that, once built, nuclear power stations have low running costs and therefore come first in the merit order of use.

The Government intervened short-term in the free market by paying subsidies to the coal industry so that pit closures could be phased in an orderly way. But the strength of the principle of market competition between fuels on a strictly commercial basis is emphasised by protests at Government intervention in the industries' key decisions.

The fundamental question facing us is whether the traditional policy of commercial competition between fuel supply industries, together with a nuclear programme decided directly by the Government, can be confidently accepted as ensuring that consumers will be in future able to satisfy their energy needs at a cost they can afford with some guarantee of long-term security.

This question has to be answered now in a situation different from that of the 1950s and 1960s. Not only has the cost of imported oil risen four-fold, there is also a general realisation that world reserves of oil and natural gas could be largely exhausted in a few decades and a consequent anxiety about how energy needs are to be met at all in the long term.

Prospects under existing policy

If energy policy continues to operate on traditional lines, gas will remain relatively cheap until the bulk of offshore resources has been depleted. Coal and nuclear electricity will remain relatively expensive. The price of oil is more uncertain, depending on OPEC's ability to maintain an effective cartel before the time when reserves approach exhaustion. Electricity will remain expensive unless it is generated from cheap gas.

The broad effects of such a pattern of prices on energy supply and use can be roughly foreseen. Gas will gain an expanding market. Coal development will be in some jeopardy because of its weak market position. As long as cheap gas is widely available, conservation measures and the development of solar, wave, tidal and wind energy will at best proceed slowly (unless technical break throughs give them a large new cost advantage).

Offshore reserves of gas and oil will be depleted rapidly, even if the OPEC oil price weakens, unless the Government uses its powers to hold up development and order cut-backs in production.

Consumers who have been badly hit by the rise in fuel prices—notably families in all-electric homes—will only be able to remedy their position if they can convert to gas or adopt very effective conservation measures.

It can be argued that heavy reliance on offshore gas and oil, combined with Government-supported nuclear development and a coal industry subject to a strong competitive pressure will provide the cheapest fuel for energy users and that long-term security will soon be provided by cheap, safe and reliable nuclear power. But there are risks and dangers inherent in such a traditional policy. A more planned, long-term approach may now be needed.

The problem for consumers

One notable improvement since the war for most families has been that they became able to afford to heat their whole house to a reasonable standard and to have an adequate supply of hot water. But the recent sudden rise in fuel prices, particularly of electricity, has recreated the problem of fuel poverty for many families. The short-run expedient of means-tested assistance such as extra heating additions to supplementary benefit, is a very imperfect remedy.

One problem about reliance on a free market system is that unanticipated sharp changes in prices which can occur are very harmful for domestic consumers. In an era of high prices for fuels other than gas with risks of long-term scarcity driving prices even higher, a major, planned programme aimed at building new houses to a much higher standard of insulation and appliances and at improving standards in existing houses would provide the best long-term guarantee that families will be able to afford a reasonable standard of warmth.

Much higher standards of energy use in the home are already achieved in countries with colder climates. From the standpoint of household warmth a rise in the cost of fuel has the same effect as a worsening of the climate. Our winters have already become 'colder' in this sense, and there is a long-term risk that they will get 'colder' still. In these circumstances the machinery of building standards, improvement grants, etc, could be used for a long-term programme to improve standards of energy use in the home. The cost of such a programme would be balanced by savings on investment to expand fuel supply.

The problem of security of supply

Traditional policy allows no real planning nor long-term security other than the nuclear power programme. But nuclear power itself involves significant risks—the likelihood of uranium scarcity and uncertainty over the economics and acceptability of high-temperature or breeder reactors which could spin out the availability of nuclear fuel.

The traditional maxim has been that in the face of long-term uncertainty we should keep as many options open as possible. In practice this is difficult because changes in the pattern of energy supply and use are slow and costly to undertake. Annual investment in energy supply in Britain amounts to some £2 billion a year, about one-tenth of total national investment for all purposes. Most of this investment is in long-life installations which imply long-term commitments. For example, a power station built now would be meeting energy needs in the 1990s and after.

The immediate options in primary fuel supply are coal, oil, gas and nuclear. To keep the coal option open requires steady long-term investment in new pits since coal supply cannot be rapidly expanded to meet sudden needs. The main way to keep oil and natural gas options open must be to restrict depletion and maintain offshore reserves. The nuclear option is more flexible but even here the installation of power stations and fuel processing plants is a lengthy business.

A policy for long-term security of fuel supply also requires development of renewable sources such as solar, wave and tidal power.

The pattern of energy use

Planned development of sources of primary fuel supply would conflict with the traditional policy of free market decisions on energy use. Investment in higher coal production, for example, is self-defeating if there is no market for the extra coal because users have switched from electricity to gas.

The Government will be able to influence the choice between primary fuels to some degree by its decisions on which of the 10–20 new power stations needed between now and 1990 will be nuclear or coal-fired. It can also informally influence the choice of fuel for electricity supply by asking the CEGB to modify the normal merit order of use. But at present it has almost no influence on the key choices by energy users between coal, gas, oil and electricity.

In order to implement planned, long-term development of fuel supply the Government would need new means of influencing the pattern of fuel use so that demand and supply for each fuel could be kept in reasonable balance. As far as fuel input to electricity generation is concerned, the Government might exercise direct administrative control and over-ride the merit order of use in the interests of national fuel balance.

But it would not be practical to determine the entire pattern of fuel use by administrative direction. A more convenient and acceptable method in the long term would be to fix relative prices of the various fuels in different uses by means of differential taxes and subsidies so as to guide choices by fuel users in conformity with long-term objectives.

Implications of a planned energy policy

It has been seen that a planned energy policy would involve deciding which fuels should be used, and by whom and how they should be used, just as much as deciding how the supply industries should develop. Decisions on energy strategy would therefore directly affect users of energy as well as fuel industries and the manufacturing industries which supply plant and equipment to the energy sector.

This would make it all the more important to end the tradition of secrecy in discussion and decision-making on energy policy. The Government would have to be made fully accountable for its policy choices and for the use of powers to implement them.

To achieve real accountability, the Government must present issues and options in advance of reaching decisions so that they can be discussed by all the interests involved, including management unions and energy users. The Government must also make regular reports on the development of energy policy as a whole and on forecast long-term prospects. A planned energy policy should therefore involve not only new Government powers to promote development of supply and to regulate prices, but also formal arrangements to ensure public discussion and joint decision-making on the key planning issues.

TUC fuel and power policy (TUC Fuel and Power Industries Committee)

Introduction

1 The TUC Fuel and Power Industries Committee which has representatives from all affiliated unions in the energy producing sectors is the body responsible for progressing Congress policy on energy matters. The TUC has consistently argued for a co-ordinated approach to energy production and indeed was the first national body to put forward proposals for an integrated national energy policy over the medium term. The TUC's 1973 *Statement on Fuel and Power Policy* pointed out that the discovery of large quantities of oil in the North Sea, the rapid exploitation of natural gas, the setbacks in the nuclear programme, the reversal of the decline in the coal industry and the general rise in the prices of fuels, all highlighted a clear and imperative need for a new national fuel and power policy in the UK. The TUC conducted a further review of its fuel and power policy in December 1974 following the dramatic rise in oil prices at the end of 1973 and the choice of reactor type in the summer of 1974. The following paragraphs summarise this latest review of Fuel and Power Policy by the TUC Fuel and Power Industries Committee and present the arguments for a co-ordinated national fuel and power policy.

Requirements of fuel and power policy

2 The time horizon for policy should be at least 10 years. A shorter period would not leave sufficient time for investment decisions to come to fruition. The policy has also to be set against the perspective of a longer view since a 10-year investment programme would largely determine the pattern of the energy production for a long period afterwards.

3 The policy set out in this document tries to avoid the major error of previous attempts by successive Governments at formulating energy policies. This has been to concentrate on the most promising fuel industry and encourage it to expand as rapidly as possible, with insufficient regard for the effects on other fuel industries. The TUC policy, while intended to be flexible and hence not upset by modest deviations from the individual fuel targets set for 1985 and beyond must be viewed as an interlocking set of policies for the component industries of the fuel and power sector, the effects of which will be fully realised in the 1980s. It is also important that the pricing and investment policies of the various industries are made compatible with the recommended production targets. The role of prices in resource allocation—however imperfect they are in this respect—should not be neglected, especially in relation to the energy sector which affects all other sectors of the economy. Apart from the need for internal flexibility within the policy, the planned targets for individual industries and for the energy sector, will need to be revised regularly, and in a major way every five years. The aim must be to introduce interventionist planning rather than to have an immutable plan.

4 There are a number of key issues which must be considered when planning the development of the fuel and power sector. Most important among these are:

(a) The adequacy of supplies (this is a separate point from the equally relevant question of security of supplies, though the two are often confused);

(b) the foreign exchange cost of the various fuels;

(c) the retention of sufficient flexibility (as far as this is obtainable at reasonable cost) to take account of the considerable economic and technological uncertainties affecting the fuel and power industries;

(d) manpower and social (including environmental) considerations.

Current prospects

5 The following paragraphs review current prospects in each of the four main sources of primary energy.

Oil

6 The disruption caused by the near five-fold increase in oil prices and the cutbacks in production have illustrated the vulnerability to the economy created by the rapid run down of the coal industry. In the North Sea production of oil could rise to over 100 million tons annually by 1980. Public control over the oil resources will be achieved through the revenue raising provisions in the Oil Taxation Act; through the actions of the British National Oil Corporation (BNOC) as an operator in the North Sea; through 51 per cent Government participation with the oil companies and through the provisions of the Petroleum and Submarine Pipelines Act. A further aspect of the emerging UK oil industry deserves emphasis. The Government has already acted towards ensuring the basis of orderly trade unionism offshore: a draft understanding about access to rigs has been circulated; employment legislation—effective onshore— is being extended this month. In continuing conjunction with employers and unions, these efforts should be intensified.

Coal

7 The TUC and the unions in the coal industry have welcomed and supported the Government's endorsement of the NCB's plans to raise coal production to 150 million tons a year by means of a £1,400million investment programme. Annual discoveries of new coal reserves continue to outstrip annual rates of depletion.

Nuclear power

8 The Committee strongly support the 4,000 MW SGHWR programme and recognise that resources in the nuclear industry must be deployed so as to give this programme priority. The trade union movement must be involved in the nuclear construction programme through representation on the Nuclear Power Advisory Board and the National Nuclear Corporation.

Gas

9 The level and structure of gas tariffs have contributed to a rapid rise in the sales of gas. Proven reserves can sustain a flow of 5,000 million cubic feet a day (mcfd) up to the late 1980s and access to the Frigg gas field will raise this total to 6000 mcfd which is twice the 1973 total.

Energy outlook to 1985

10 The demand for energy can be broadly estimated by looking at the prospects for growth in the economy as a whole and at the relationship between economic growth and the consumption of energy, which is called the energy coefficient. With economic growth averaging 3 per cent a year between 1974 and 1985 and with an energy coefficient of between 0·5 and 0·9, the total demand for energy would increase to a level lying between 410 and 460 million tons of coal equivalent in 1985. The Table shows the TUC proposals for supplying this demand from each of the main sources of primary energy.

11 The *coal* figure of 150 million tons output has already been given official support, and this support must manifest itself in action on the investment programme to ensure that there is sufficient capacity to enable the realisation of the target figure. The *nuclear* output of 31 mtce takes account of the nuclear capacity which is already producing electricity the 5,000MW AGR programme the 4,000MW SGHWR programme and further nuclear orders of 1,500MW in 1979. Natural gas from existing gas fields and from Frigg can supply 6,000 mcfd or 80 mtce in a year, but at this rate of depletion, supplies would be falling away in the 1990s. Reserves of *oil* in the North Sea would allow a depletion rate of 150 million tons a year, although the latest official estimates based on the progress of development of oilfields suggests that production in 1980 will lie in the range of 95-115 million tons. Nevertheless the UK could still be producing energy surplus to its own requirements in the early 1980s.

Prospects for 1995

12 If the projected depletion rates of oil and gas were at the top end of the range shown in Table 1 a substantial shortfall of energy, up to 270 mtce, could develop by the late 1990s. One consequence of this could be the construction of up to 100 nuclear power stations each of 1,300MW capacity in the period 1985-1995, or the return to importing substantial amounts of oil.

TUC recommendations

13 Table 1 compares the TUC's proposals for the prospects for energy production in 1985 with those which were presented by the Department of Energy to the Energy Resources Sub-Committee of the Select Committee on Science and Technology in March 1975. The forecasts were the results of the Department's 1974 Energy Review. The comparison shows that the Government are postulating a lower figure for coal production and a higher figure for gas depletion than the TUC, though the Government state that the 150 million ton figure for coal would be reached if optimum

use were made of existing capacity and if all new projects were brought in efficiently and on time.

UK energy demand and supply: 1985
(million tons coal equivalent)

UK Production	Present policies (1)	TUC Recommendations (a) Average economic growth (2)	TUC Recommendations (b) High economic growth (3)
Coal	135	150	150
Nuclear —	40	30	30
Hydro —		2	2
Gas	80	60	60
Oil	170–255	180	240
Total Production	425–490	420	480
Total Consumption	410	410	460

14 The TUC is also arguing for a rate of depletion of North Sea oil lower than the top end of the Government projection of 255 mtce in 1985 if that oil surplus were to be exported in primary form. The proposed rate of depletion is illustrated in column 2 of the table when economic growth would be expected to be running at no more than its trend rate. However, the availability of North Sea oil will permit higher economic growth in the UK. The TUC therefore considers that a high economic growth strategy should be adopted in the 1980s. This would be accomplished through bringing about structural changes in the economy to enable it to move into surplus on the non oil current account through specialisation in the high value added products associated with oil. If this strategy were adopted then the pattern of demand for and supply of energy in 1985 would be broadly as illustrated in column 3 of the Table. The details for each of the energy industries are as follows.

Coal industry

15 The Government's commitment to the coal industry has been widely welcomed. There are two broad areas where policy has to be developed. The first is in the field of technology and the NCB's research programme was given backing in the Tripartite Examination. It is clear that if major new techniques such as the fluidised bed boiler and liquifaction, are perfected, then coal will be used not simply as a primary energy input but also as an input into the petrochemicals industry. It will also mean that coal can be substituted directly for oil oil and gas, and this is particularly important if oil and gas are seen as premium fuels.

16 Secondly more needs to be done immediately to guarantee the market for coal. The building up of coal stocks, financed by Government is welcome, but action needs to be taken on coal imports. Consideration needs to be given to long term contracts being drawn up between the NCB and major users such as CEGB, the SSEB, and the British Steel Corporation. The commissioning of new coal fired power stations at present would mean that when they come on stream some oil fired power stations will have to be scrapped prematurely. However, it is possible to convert some oil fired stations to coal fired capacity. This will also mean shorter lead times than the construction of new power stations.

Nuclear power

17 The limited SGHWR programme has been welcomed by the Committee. However, the Committee equally recognise that this is not an area where there can be undue delay given the very long lead times involved. It is essential that the SGHWR programme is completed as quickly as possible so that a fuller programme can be embarked on after 1978. It is essential that the appropriate trade unions are fully involved in the nuclear programme, and the Committee have pressed for representation on the National Nuclear Corporation and on the Nuclear Power Advisory Board.

18 A second area of delay could be associated with the reactor to follow the SGHWR. The TUC recognises that the role which will be required of the Fast Breeder Reactor (FBR) in the late 1990s demands an intensification of the research effort.

Gas industry

19 The Committee recommended that gas should be seen as a premium fuel and the rate of depletion should reflect this. The premium uses of gas are for household heating and cooking and for some specialised industrial uses, for example, glass manufacture. However, gas should not be used generally in industry, neither should it be used as a power station fuel input. One difficulty is that on present arrangements gas will be landed together with North Sea oil. This problem will in part be solved by a lower rate of depletion of North Sea oil, but it will also have to be tackled by the renegotiation of the British Gas Corporation's contracts with the oil producers. This should be part of the renegotiation of licences promised by the Government. The TUC recognises that if gas is to be a premium fuel the level and structure of its prices will have to reflect this. Further price

increases should therefore be tilted away from domestic towards industrial users of gas who already enjoy a substantially lower price.

Oil industry

20 The resolution on energy carried at the 1974 Congress called for the public ownership of North Sea oil. The TUC has taken the view that the Government's proposals as outlined in their White Paper (Cmnd. 5696) will be an effective way of achieving public ownership. The TUC believes it is important that the BNOC is given the fullest opportunity to exploit its position and it should be adequately represented on the operating committees of consortia holding North Sea licences. But BNOC must not only acquire majority holdings in oil fields, it must develop its operations in activities such as refining and distribution and the petrochemicals industry. The TUC attaches importance to a clear expression of the Government majority stake in British Petroleum Ltd. The Government should also use planning agreements in an active way to encourage the existing oil companies to expand the petrochemicals industry. It is vital that there is a planned approach to avoid the dangers of either over or under capacity.

21 There also need to be developments in the supply of equipment for oil exploration and exploitation. The public ownership of construction sites for oil platforms will be a valuable base for planning in this area but more needs to be done. The proportion of the market for offshore oil supplies supplied by UK industry has increased from 25–30 per cent in 1973 to about 40 per cent in 1974. Nevertheless, many orders for equipment are still being placed overseas. Later this month the Oil Industry Liaison Committee (which is chaired by the Minister of State for Energy and embodies platform construction interests) will discuss the Energy Department's updated forecast for platform orders based on the series of discussions Energy Ministers have had with oil companies over recent months. There can be no doubt about the urgent need for platform orders to be directed towards such sites as Graythorpe and Methil, already active, as well as those in Scotland which still await their first contract. The 'one-off' nature of platforms is understood but construction in anticipation of demand should not be ruled out. Planning agreements and public ownership should be actively used to ensure that UK industry can meet future demands in this area. One possibility is for the BNOC to acquire some specific supplying interests. The TUC will be examining this area in more detail.

Policy co-ordination

22 In the 1973 Fuel and Power Policy statement the TUC urged the Government to set up a Fuel and Power Board to ensure the co-ordinated development of the various energy industries, with overall planning responsibility to the Energy Secretary. It is not envisaged that the Board would interfere in the day to day running of the industries. A similar idea is expressed in the 1974 Congress resolution which called for a Standing Commission to investigate the co-ordination of the energy industries. This document has shown the need for co-ordination if the problems of energy supply and demand are to be met. A co-ordinated aproach is also needed for pricing and investment.

23 It has been recognised that a Fuel and Power Board would need to possess the power to require both the private and the public energy industries to conform to the *outlines* of the co-ordinated plan for the sector. A Board with purely advisory powers would probably have very little effect on policy decisions.

Summary of main recommendations

(i) The major coal users, the CEGB, SSEB, and BSC, should enter into long term contracts with the NCB;

(ii) Research into new coal technology should be intensified;

(iii) The coal burn of the CEGB and the SSEB should be increased by the construction of new coal-fired power stations and the conversion of oil to coal firing when and where possible;

(iv) Both the CEGB and General Electric Company are showing a lack of urgency on the nuclear programme, and there is a case for bringing forward the date at which the programme can be expanded;

(v) There should be trade union representation on the National Nuclear Corporation and priority should be given to the SGHWR programme;

(vi) There should be trade union representation on the Nuclear Power Advisory Board to advise as a matter of urgency on the post 1978 nuclear programme;

(vii) A major effort should be made on developing the FBR;

(viii) Natural gas should be considered as a premium fuel and its price and rate of depletion should reflect this. Any further price increases should be tilted to industrial users;

(ix) More North Sea oil should be used as an input into an expanded UK petrochemical industry rather than as a primary fuel input;

(x) The BNOC should use its remit to develop a refining, distributive and petrochemical base;

(xi) A clear impression of the Government majority stake in British Petroleum is needed;

(xii) The establishment of orderly trade union organisation offshore should be actively pursued;

(xiii) Maximum use should be made of planning agreements and public enterprise to develop planned growth of oil equipment and petrochemical industries. Companies should also reveal information on the likely rate of production from UK sources;

(xiv) Measures to overcome the short-term dearth of platform orders should be actively pursued;

(xv) Exports of oil should be in the form of high added value refined products rather than crude oil;

(xvi) The Government should give attention to ways of achieving greater co-ordination of energy policy, through a National Fuel and Power Board or a Standing Commission;

(xvii) The Government should be asked to set in hand the preparation of its own 'jigsaw' of the pattern of total energy assessments for the mid 1980s.

Energy policy—a trade union strategy (National Union of Mineworkers; Electrical Power Engineers Association; National Association of Colliery Overmen, Deputies and Shotfirers)

The need for a more coherent and carefully planned approach to energy policy is well established and scarcely needs repeating. It was evident long before the Middle East oil crisis that the traditional haphazard methods could not guarantee the availability of essential fuel supplies and it was with this in mind that the trade union movement set about the preparation of a policy statement on energy which was adopted by Congress and published in 1973. Although much has changed since then, the principles underlying the TUC approach to energy policy at that time remain valid today.

Indeed, with the creation of a British National Oil Corporation to exploit our North Sea oil reserves for the benefit of the British people as a whole, the situation exists now more clearly than ever for the various interests involved in the energy sector jointly to agree on the outlines of a new approach which will reflect not merely the various parochial interests at stake, but also the points of common concern to which the TUC has drawn attention over the past few years.

One such point brought out in the 1973 statement was that;

> 'The energy sector is not a typical industrial sector. Energy costs enter the costs of production of all other industries and all households and commercial premises consume energy. Energy is thus part of the economic infra-structure of the whole economy and as such adequacy and security of supply and obtaining energy at the lowest reasible resource cost is essential to the general efficiency of the economy.'

The policy framework

With this in mind, the need to produce an agreed framework for reaching policy decisions becomes crucial and there are three clearly identifiable elements which such a framework must contain.

1 Firstly it is imperative that corporate planning in each of the different fuel industries should be conducted on a fully tripartite basis. The example of the Coal Industry Examination in 1974 provides an excellent model for this. It has produced visible advantages for the coal industry and its employees, at a time when the deepening industrial recession has provoked a review of capital expenditure programmes elsewhere throughout the public sector, and structured planning on tripartite lines in other energy industries would provide a much needed opportunity for trade union representatives to make their voice heard in the formulation of decisions vitally affecting the interests of workers.

2 In addition to tripartite planning, provision must be made within the policy framework for full public debate. It is possible that without such provision, uninformed public opinion could for example prove a major obstacle to the expansion of nuclear power, as it has in other countries, and the responsibility therefore lies with the Department of Energy to encourage widespread discussion of policy options on the basis of accurate and freely obtainable information. One way in which this might be achieved would be by the publication of regular White Papers by the Department of Energy describing current market forecasts, basic policy assumptions and the progress being made towards stated objectives. A White Paper of this kind could appropriately by produced once every three years and there could perhaps be a shorter annual review in the intervening period.

3 Finally it should be recognised that it is not sufficient to lay down planning procedures for individual fuel industries unless these bear some relation to the integrated development of the energy sector as a whole. Thus the emerging form for consultations between the different energy industries must be instituted on a more permanent basis, with trade unions continuing to play a full part and under the chairmanship of the Secretary of State for Energy, to determine the basic pattern of long-term energy supply.

The nature of fuel resouces

With such a policy framework established, and following the speedy implementation of the

Plowden recommendations on the structure of the electricity supply industry, agreement can more readily be reached on the likely future movements in the price and supply pattern for each of our fuel resources. This will also involve taking note of the intrinsic qualities of the fuels themselves and of the priority to be given to balance of payments and security of supply considerations, but certain broad characterisations in this area are not in dispute.

On existing forecasts it is clear that there is a strong likelihood of an energy gap re-opening sometime in the 1990s unless direct action is taken to increase investment within the energy sector to prevent this happening. That this situation should exist in a country so apparently well equipped with energy resources may seem paradoxical but this apparent abundance must not be used as an excuse for complacency, still less for the over-rapid exploitation of fossil-fuel resources.

Our reserves of coal are very considerable indeed and the industry is currently able to produce at a level above that of the present depressed demand. The *Plan for Coal* has already set out objectives for that industry and it is clear that with the current price advantage that coal enjoys over oil then, in the likely event of this being maintained, coal must remain by far the main fossil fuel source for electricity within the UK.

Total reserves of oil and gas remain unknown, but even when estimates have been made on the most optimistic basis there is no suggestion that the life span of these fuels will be of the same order as for coal. It is also the case that, although coal is primarily an energy source, hydro-carbon fuels are increasingly valuable as a feedstuff to the chemical industry quite apart from being premium fuels with particular attraction to specialised markets in their energy use. Certainly in the 1990s and succeeding decades the real price of hydrocarbon fuels seems bound to rise as world shortages appear and it would be foolish to ignore this factor in our production and use of energy in the intervening period. There will also be balance of payments pressure on the United Kingdom to export these fuels in the interests of strengthing the whole footing of the economy.

These and other limitations on hydrocarbon resources re-emphasise the importance not only of coal but also of the British nuclear industry. At the moment nuclear power is our cheapest single source of energy and it is likely to continue to be so. Because of its high capital cost it is particularly suited for meeting a substantial part of base load demand for electricity. While the emphasis on the rate of development of additional nuclear capacity in the UK in the short term may not be as fast as elsewhere in the industrial world, in the long term it seems unavoidable that it will have to meet an increasing share of UK energy demands.

Fuel mix in the supply industry

The particular demand problems in the supply industry at the moment mean there is a considerable, and understandable, reluctance on behalf of the generating boards to order new plant. Forecasting electricity demand has always been a hazardous occupation but the pessimistic turn of current *medium* term forecasts would not seem compatible with the likely energy demands of the UK given even a modest rate of consistent economic growth. For example, a failure to invest in new plant construction for the next few years which would be the logical outcome of unduly pessimistic assessments about the future would have immensely serious consequences.

1 It would mean the annihilation of any UK ability to manufacture nuclear power stations, leaving us wholly dependent on imported technology and plant (and *de facto* safety standards) thereby undermining or reversing the policy of the Government as set out in the 1974 White Paper on reactor choice (Cmnd. 5995).

2 It would seriously damage the ability of the coal industry to fulfill its obligations under the tripartite plan by jeopardising the investment programme which depends in large measure on extra coal production being converted into cheap electricity in modern high-efficiency plant.

3 There would be considerable damage to all the plant manufacturers and firms who provide plant of this kind to the industry in this country. The extent to which some of these would remain viable must be called in question.

These three points need to be clearly understood. They show the very hazardous consequences of adopting long-term pessimistic forecasts as a result of essentially short-term pressures. The extent of pessimism has principally arisen as a direct consequence of the price advantage which gas enjoys over electricity in many of its competing markets at the moment. However the consequences of allowing this price advantage to continue without further consideration both of the state of our long term gas reserves and of the clearly understood need to retain viable electricity and coal industries could be as serious as is indicated above. In this context it is necessary to take full account of the EEC forecast that the electricity supply industry will in the long-run continue to increase its share of the total energy market.

Clearly in the situation where there will be alternative uses for the heavy fuel oil currently burnt in power stations, or at least for much of it, and where the gas used to generate electricity at the moment should be priced at a higher level, then the use of these two fuels by the generating boards in this country seems unlikely to increase. This means from the point of view of electricity generation that there will need to be a balanced programme for the construction of new nuclear and coal fired capacity beginning with the proposed nuclear stations at Sizewell and Torness and the proposed coal station at Drax.

Government policy on the construction of the Drax power station has been that the power station to be built there should be timed to coincide with the development of the Selby Coalfield. This was accepted by both the CEGB and the NCB. Now that the Secretary of State has given approval to the planning application for mining at Selby this requirement has been met. The generating board is, however, now reluctant to order new plant which it does not think will yet be required to meet its statutory obligations for the economic generation of electricity. Under these circumstances, in view of the importance that the Drax power station quite clearly has both to the economic justification of the Selby project and in terms of the future recruitment and morale within the coal-mining industry, it would seem appropriate that the Secretary of State for Energy should seek to reach agreement with the CEGB for the construction of this power station in the near future.

In the case of the nuclear industry the situation is somewhat different. The SGHWR programme, initiated by the present Government in 1974 and fully supported by the TUC, has required particular care at the design stage which has required the original timetable to be revised. Firmer long-term dates are now being provided by the NPC and by the generating boards for the likely timetable for the first reactors on each of these stations to be on full power. It is important that the Government should give full support to the completion of the programme within the revised timetable so that there will be an adequate basis for subsequent nuclear re-ordering. This would provide for an independent British nuclear industry able to produce considerable extra generating plant during the probably crucial period of the 1990s.

These proposals could, however, be effectively jeopardised were the gas industry to be encouraged to follow policies of all out expansion whereby its advantages as a monopoly buyer of gas from the companies exploring the North Sea were to be handed over directly to the consumer in lower prices. The TUC has made clear its policy that gas as a premium fuel should in fact be at a premium price and it is to be hoped that the gas industry will now take the necessary steps to ensure that all its prices meet these criteria. Additionally now that nationalised industries are to return to a system of financial objectives it would seem appropriate for the Government to examine whether the Gas Corporation ought to show a higher rate of return than the electricity industry. Otherwise dislocations in employment in the energy industries are likely to follow the over-rapid expansion and subsequent necessary contraction of the gas industry share of the energy market.

In the oil industry some action to reduce overall demand and minimise import dependance has already been taken. One such measure is the proposal within the EEC to set an import barrier in oil at prices below \$7 per barrel but there are legitimate grounds to doubt this will have the desired effect of protecting investment in indigenous fuel production. Investment in North Sea oil fields, and in coal expansion under the ten-year plan, has been committed on the basis of oil prices at around \$11 per barrel. Any figure much below \$9 could therefore prove useless as a minimum import price designed to eliminate threats to continue exploration and development of UK fuel resources and it is to be hoped that the Government has not finally committed itself to accepting the \$7 figure.

It must be hoped also that the Government's position on oil depletion is not yet firm. Powers have been taken under the Petroleum and Submarine Pipelines Act to enable the Secretary of State to exercise a considerable measure of control over depletion but the indications are that these will not be fully used until production in the North Sea has reached its peak. By this time oil may have made many inroads into non-premium markets and though there are of course other factors to be weighed including the overall needs of the economy and the implications for the financing of further North Sea development, on energy grounds alone careful control of offshore oil depletion is a foremost priority.

Long term consideraions

The proposals in the previous paragraphs relate to the need to establish the correct energy mix between different fuels while being able to provide the necessary resources to deal with the potential energy gap in the 1990s. They do not deal with the possibility of long term changes in the level of demand as a result of direct conservation of energy by consumers nor with the development of new alternative energy sources. Nonetheless both these will clearly be major developments between now and the 1990s.

With the advantage of hindsight, it is self evident that the allocation of resources to build four different prototype nuclear reactors simultaneously was a large scale misuse of resources and the minimal amount of research on coal utilisation during this period serves to highlight this. Nonetheless it is important that the energy sector does not now neglect either original research or potential development of existing technology when it can be shown that savings in both economic and resource cost are obtainable. We can be sure that none of our major competitors will neglect this. In addition to the continuing support of the SGHWR programme beyond the original prototype, it would be sensible to seek to expand the development of fluidised combustion of coal, particularly in industrial boilers. In the long run research on fusion development and on the liquefaction and gasification of coal would seem to be of major importance.

Similarly the general agreement that new buildings should be designed with a greater regard for their energy requirements should be given specific form as soon as possible. Although by its nature this is not something which can have immediate impact it is necessary to start immediately in order that the slow process of change can begin. It seems much more important for the Government to devote its energy to promoting fundamental change of this kind rather than to the short-term 'Save It' campaign.

The immediate issues

Growing cynicism towards the 'Save It' campaign is one issue of immediate concern in the energy sector. There are many other such issues and some of those which were raised during the joint consultations between the coal and electricity industries on February 20 are already receiving attention. It is nevertheless useful to remind ourselves of these as the prospects for the future development of the sector rest on the satisfactory resolution of the immediate issues within it.

The problem of coal imports is among those receiving attention. Although the generating board's cutback on imports in 1976/77 is to be welcomed, there can be no cause for complacency until steam coal imports have been eliminated. At a time when British coal is being put to stock, foreign imports, particularly at higher prices, are without justification. Instead we should recognise that coal expansion on a worldwide scale offers excellent opportunities for exports, particularly within the European Community, and these should be seized as conditions permit.

One by-product of the temporarily depressed level of electricity consumption is that a certain amount of power station conversion would be technically possible without putting at risk the supply industry's ability to meet maximum demand. Against this background it is necessary to examine the economics of the specific proposals put by the coal industry to the generating board.

At the February 20 meeting specific attention was drawn to problems in both Scotland and Wales. In the case of Wales the generating board and the Coal Board appeared to have already reached satisfactory solutions with the decision to rebuild the Aberthaw B set so that the CEGB capacity in Wales would be more capable of burning NCB output. It is to be hoped that the temporary breakdown on May 5 will in no way have altered the principle behind this agreement. In Scotland the quality and quantity of new generating plant being commissioned or under construction contributes to the problems faced by the coal industry there during the present recession. The protracted negotiations over the price of coal supplies to the Longannet Power Station between the SSEB and the NCB are also a major contribution to uncertainty. It is to be hoped that it will be possible for the authorities soon to agree if necessary within the framework of the tripartite working party which met on April 2, the appropriate price levels which will continue to permit Longannet to be an essential part of the South of Scotland base load capacity.

Policy issues requiring immediate attention in the area of offshore oil development are legion. Among them are the conditions to be attached to the next round of licensing and the extent to which this must be used as a lever to secure participation in existing licences and new orders for British firms operating in the offshore supply industry. No simple solution to these problems is available but it is now evident that without a more vigorous and positive approach by the Government the objectives of the voluntary approach so far adopted could be seriously undermined.

Whatever the immediate prospects for the offshore oil and gas industries it is clear that electricity will have a major role to play in filling the energy gap in the 1990s, but only through being at that stage firmly based on the coal and nuclear industries with the nuclear power share gradually increasing as coal conversion replaces dwindling supplies of hydrocarbon fuels. To reach that position the appropriate investment and policy decisions must be taken now and it is in this context that consultations between different fuel industries should be seen.

Energy policy—the case for integration and expansion (Amalgamated Union of Engineering Workers)

AUEW policy conference decision:

'This National Conference instructs the National Executive Council to press the Government to bring Off-shore Energy Resources under public control and as a first step:

1 Majority participation in all future licences.

2 Negotiate majority participation in existing licences.

3 Take powers to control the pace of depletion.

That an integrated fuel policy should be developed so as to take into account the part to be played by Oil, Gas, Coal, Electricity and Nuclear Energy in the next 25 years. That revenue from Off-shore Oil must be used to create self-generating industries in the Development Area.

We consider that it is necessary the BNOC should become the dominant company in the Seas surrounding the UK and that it take into complete ownership BP as a basis for expanding in all fields of oil activity in our country.'

The United Kingdom is one of the world's leading manufacturing nations and exporter of manufactured goods, and is consequently a high energy-consuming nation. We must approach our energy policy on the assumption of an increasing standard of life and the need for technological advance, and the maximum development of an integrated energy policy which will eliminate wasteful competition between energy producers, whilst permitting the best use of our natural resources and protecting the environment.

We believe that in recent years wasteful competitive advertising in coal, gas, electricity and oil has been apparent and has reflected temporary advantages for panic policies which have been totally unplanned.

In the Electricity Supply Industry we recognise the role of nuclear energy as a vital programme, and welcome the Government decision to use British steam generating heavy water reactors. We believe this programme should be approached with enthusiasm and urgency and will reflect considerably on the Manufacturing Nuclear Industry and the technical development of SGHWR.

The Nuclear Industry requires such stimulation and design encouragement with potential sales advantages overseas. In this connection, we believe there should be an early decision by the Government to construct at Caithness a large Fast Breeder Reactor Power Station, which would itself be of major importance in developing confidence in the Nuclear Industry and contributing to the maintenance of a leading position in this scientifically based high technological field.

We further welcome the announcement by the Government of an initial investment of £1 million on development and research in the field of wave power and also in the solar power field. We believe new investment is necessary on the best uses of new techniques of consuming the vast resources of coal and the alternative uses of these resources. It is our view that coal should remain the main energy source for the Electricity Supply Industry in this era, and we welcome the statement by the Electricity Generating Board at the earlier conference in this connection.

We believe a programme organised jointly by the Electricity Supply Industry and the National Coal Board is essential.

North Sea oil and gas

The energy scene has not fully absorbed the impact of the discovery and development of North Sea gas and oil. Many speculative estimates on the impact on other energy resources and the possible extension of use of oil and gas into electricity generating would be, in our view, wasteful of those resources. We welcome the creation of the BNOC and the Government must display the same determination for ensuring its development as a fully integrated oil company. We believe British Petroleum should be taken over by

BNOC as a first step towards an integrated expansion policy.

It would be easy to be tempted to over-exploit these resources for purely economic short-term advantages, but we believe that a planned conservation policy must be the overriding factor with the enormous long-term advantage on the balance of payments. In this connection, the extraction rates should be strictly regulated to generate United Kingdom self-sufficiency.

Integration and expansion

We believe, as a matter of urgency, the Government and the industries concerned should establish a National Fuel Co-ordinating Council, whose terms of reference would include forecasting of energy requirements, examination of priority development in the four main fields under a central energy policy. Such a council should integrate its plans to link up with the manufacturing industries in electrical and power engineering, and be required to establish direct links with the industries who supply equipment and exploitation resources. We believe it is extremely important to establish a higher level of confidence and continuity in our home-based industries.

We further suggest that wasteful advertising and unnecessary competition should be eliminated and such policies closely examined.

We conclude by summarising that a relatively small but influencial Energy Council or Board, if established under the control of the Minister and accountable to Parliament, would be of enormous importance in developing a higher level of confidence on the future allocation and usage of our resources, with particular emphasis on conservation, and a distinct advantage on the export programmes of both resources and technology, and would certainly influence a much sounder based manufacturing policy which would be of considerable assistance in the Engineering Industry.

Statement on energy policy (Electrical, Electronic, Telecommunications and Plumbing Unions)

We are as conscious as anyone of the weakness of long-term predictions and of the near inevitablility of their inaccuracy. Accordingly, we are extremely concerned that there should not be an over-dependence or over-emphasis on a particular energy source. We adopt this policy for strategic defence reasons, for reasons of balance of payments and because of the need of Britain's industry for secure cheap energy.

Assuming the historic energy demand growth rate of 1·8 per cent, we can in this country expect an energy gap by 1990 of between 50 to 100 million tons of coal equivalent (mtce). This assumption is quite valid since, short of a radical transformation in the expectations and attitudes of the British people, a steady rate of economic growth and thereby energy consumption is necessary. Indeed, the previous historic growth rate of 1·8 per cent could well be lower than the actual rate. A failure to provide such growth, combined with an increase in productivity in the British economy, would create a built-in rise in unemployment that would be unacceptable to trade unionists and socialists alike.

So, of paramount importance must be the need to meet this energy gap—and the means by which it is met. There are several options open but most of them raise the problem of either using up vital fossil fuel reserves, or accepting conditions which are becoming more and more socially unacceptable.

While coal reserves could meet the expected deficit in energy, would there be sufficient manpower available to extract the coal? Unless we are faced with long-term heavy unemployment, work in the pits will become increasingly unattractive and will exaggerate the present position. We have nearly reached the end of the mechanisation programme in coal mines and the only way of raising, or indeed maintaining, productivity is by large scale automation. There undoubtedly is a major role for coal in the energy programme for the future, based on the TUC Fuel and Power Policy, but in itself coal will not be able to meet all our future needs. Equally, we should ask whether it is morally right to ask workers in the coal, and indeed oil, extraction industries to do unpleasant and dangerous work.

With oil and gas we have to seriously question whether it is appropriate to pour it into furnaces to provide heat and ultimately electricity when they have so many other uses, particularly as a valuable chemical feedstock. As a major industrial nation our priority should be to concentrate on raising the value added to this country by developing the petro-chemical industry. Furthermore, despite new finds of oil, we are still facing a fairly short life expectancy of reserves on both a national and international scale.

Whilst the contribution to energy production, which can be made from renewable sources, should not be overlooked it must be recognised that it will be a limited contribution, as will the saving to be made through conservation.

The remaining option, nuclear power, would seem to be the only form of energy production able to meet the required criteria. On the grounds of safety, the record of nuclear reactors, contrary to popular belief, is exemplary. There have been no recorded accidents totally attributable to the nuclear generation of power, and with the rigorous standards imposed by Britain's Nuclear Installations Inspectorate, there is no reason to expect any. However, in order to meet future requirements, the current programme of SGHWRs must be pressed forward with the utmost urgency. Also, since the supply of uranium is itself exhaustible, programmes for fast breeder reactors, which can use uranium some 60 times more efficiently than conventional systems, must be initiated and implemented.

There is much opposition to a nuclear power programme and if such opposition can make safety requirements even more rigorous, much of it is of a constructive nature.

However, there are politically motivated groups who believe that by undermining the only feasible energy policy they can create a situation whereby an energy shortage can be used as a political weapon.

To allow a situation to arise whereby such groups can dictate future energy policies to their own advantage would be totally unacceptable.

The development of the British nuclear programme is essential not only to safeguard our energy supplies but also to maintain our leading world position in nuclear technology. At stake are thousands of jobs in the power engineering and construction industries and a rapidly expanding export market. Our ability to export SGHWR technology is bound to be limited until we possess a working reactor. Accordingly, public debate, while important, must not become a delaying tactic.

Energy policy
(Transport and General Workers Union)

PREFACE

Increased oil prices created terrible problems for the ordinary people of Britain—increasing the cost of living beyond all reason and causing a massive loss of jobs.

This Conference can help us to get out of that situation and build security into an essential part of living in these cold islands.

Working-men and women want this and want a say in how it is to be done, because frankly we no longer have *blind faith* in the expert, the 'professional', and the civil servant!

1 INTRODUCTION

The discovery of oil and gas off our shores has dramatically changed, not only the country's energy prospects, but our whole economic future. Energy policy is too often thought of as an area where only 'experts' should have an opinion. Little information is made available to those working in the energy sector, let alone the public, and decisions and strategy are formulated behind closed doors.

The development of a coherent long-term energy policy is a matter of great concern to the trade unions—it is too important an issue to be left to so-called 'experts' and civil servants.

The T & GWU welcome this opportunity to put forward its views in public. The present Government policy of open discussion on the major issues of energy policy and the tripartite discussions in each industry is to be welcomed.

2 OBJECTIVES OF A PLANNED POLICY

There is an overriding need for a planned, long-term development of Britain's own energy sources. Such a planned energy policy would need to be based on a number of stated principles:

(a) to ensure that industry's needs for energy are met at a reasonable cost;

(b) to ensure that everyone can afford adequate heat and light at home;

(c) to ensure that these objectives are met on a long-term basis, taking account of the need to regulate the depletion of our reserves of oil and gas;

(d) to ensure that the evolution of policy fully involves both unions, management and consumers.

3 ENERGY PLANNING AUTHORITY

A new planning framework is long overdue for the energy sector. There must be better co-ordination between the various industries to achieve the optimum use of our resources. The Government should take early steps to set up an Energy Planning Authority (EPA). We envisage this body as being a highly practical organisation, involving management and trade unionists. Though there would be a need for professional help and advice, the emphasis would be on the involvement of those working inside the energy sector.

This body would be responsible for the evolution of energy policy. Within the broad policy framework laid down by the EPA, the individual management would be responsible for implementing this policy. This would entail a change of statutes for the nationalised bodies. At present, by law each sector must operate efficiently and economically even if this entails competition between industries in the energy sector. This militates against the evolution of any planned energy policy.

The EPA should be established by statute and the public corporations, including the BNOC, should have a duty to co-operate with it.

4 A BALANCED ENERGY POLICY

We need to use our energy reserves more wisely and to adopt a more comprehensive and balanced approach to the planning of our energy supplies. We should not be in the position of total dependence on any one energy source. The energy gap that is sometimes forecast for the 1990s should not rush us into developing a crash nuclear programme as some European countries have done. The UK has outstandingly good energy reserves, not just of off-shore oil but also of natural gas and coal, and in the future there are good prospects for nuclear power.

We support the development of the SGHWR and we have urged the Government to press forward with this programme as quickly as possible, commensurate with the need to ensure adequate safety provisions.

On present Government estimates we will be self-sufficient in oil by 1980. There is sufficient gas available in the North Sea to satisfy the essential gas market well into the next century. It is in this section that the need for a co-ordinated and planned approach is most manifest, particularly as it is primarily the responsibility of the private sector. The resources available offer a tremendous opportunity to develop industry in the regions. In particular there are important spin-off effects on the construction industry and the petro-chemical industry. As a union we have not been satisfied that there has been sufficient forward planning, as the problems of the production of oil platforms and the siting of the production facilities show.

The Government, throught the medium of the Planning Agreement, the licensing system and the BNOC, in particular, has a vital role to play in this sector and it must have a strong and dominant present in the industry. To this end the Government should consider the transfer of the present state holding of almost 70 per cent of BP to the BNOC. The activities of the BNOC should be extended, as rapidly as possible, into downstream activities, particularly in refining, distribution and in petro-chemicals. We strongly support the principle of 51 per cent state participation in the oil companies and believe this to be an important consideration for the next (fifth) round of licensing.

The revenue from this valuable national resource must assist the regeneration of British industry. Estimates have shown that (assuming no change in the levels of tax) that between 3–5 per cent GDP will be accounted for by Government 'take' from taxes on North Sea resources. If we wish to raise the level of manufacturing investment to the 5 or 6 per cent growth that is judged necessary for regeneration, then this only requires an extra $1\frac{1}{2}$–$2\frac{1}{2}$ per cent of GDP to be devoted to manufacturing investment. This would still enable part of the revenues to be devoted to establishing pensions other social security payments at a level tied to average earnings.

It has been suggested that Britain will have an energy gap by 1990. It may well be that these fears have been grossly exaggerated but nevertheless there is a need for an effective depletion policy and our sovereignty over our oil reserves must remain firmly in the hands of the UK Government and the projected EPA.

The past run-down in the coal industry has been shown to be misguided and coal has a vital role to play in the medium and even more the long-term future.

5 NEW DEAL FOR FUEL USERS

It must be a clear duty of the Government to ensure every home is equipped to use energy efficiently and that every household can afford to maintain a comfortable standard of warmth and light. The sudden sharp rise in the cost of heating and lighting has caused hardships to most consumers, but particularly to the low income families and pensioners. It is becoming increasingly clear that the prices of coal, gas and electricity are mainly determined by short-term financial objectives, since the profit and loss of each industry varies from year to year on account of fluctuations in capacity, utilisation, interest rates etc. For the foreseeable future energy will be expensive and this necessitates a new deal for fuel users. This must be a major priority for Government and discussions should start now in order that proposals can be put forward for this coming winter.

This union has proposed that the elderly and disabled should be provided with a free allowance of coal, electricity or gas. The cost of this would be little in terms of our total resources in Britain, particularly at a time when we have a large surplus of power station capacity and coal which is likely to persist until the end of the decade. But the needs of large families, the low paid, those on Social Security and unemployment payments must also be looked to. People have the right to a reasonable standard of heating and lighting in their homes, irrespective of income. A systematic Government intervention through a combination of fiscal, housing and energy policies will be needed to make that right a reality.

6 GREATER INDUSTRIAL DEMOCRACY

Planning in the energy sector must be on a fully tripartite basis. As a union we have substantial interests in the gas, electricity, open-cast coal and nuclear industries. We are the predominant union in the oil refining, distribution and extraction industry. In each of these areas we have pressed, and will continue to press for more involvement for the lay trade unionist at each level of decision-making.

Consultation and work-force unionisation is already highly developed in the public sector in Energy. But we seek to go one step further and involve our lay representatives on the boards of the public corporations. The Boards of all the industries must be re-constituted on the basis of stated TUC policy, which is for 50 per cent of the seats being occupied by trade union representatives elected from within the industries. In addition we want to see lay trade union representation on the BNOC and on the National Nuclear Corporation and the Nuclear Power Advisory Board. We will be seeking early Government action in changing the statutes of the nationalised industries in this respect and are urging our views on the present inquiry that the Government is conducting into the whole of the public sector.

Progress in the public sector in energy has not been as speedy or as innovatory as we feel it could have been. In the private sector in energy we have shown as a union that we are not afraid to make radical changes and we believe the industrial relations record here is one to be proud of. Only in one respect—that of off-shore operations—have we had cause for concern, and thanks to help from the Secretary of State for Energy, in progressing discussions with the employers, we are evolving a satisfactory solution to these problems.

7 GREEN PAPER

There must be a full public review of energy policy, covering all the relevant options and issues and the Government should set about these discussions with the relevant bodies and the trade unions as soon as possible. These discussions must be held with a view to the Government presenting a Green Paper on energy policy this year, to form the basis for a 10-year Energy Plan.

The layman's guide to the requirement for energy (Amalgamated Union of Engineering Workers—Technical, Administrative and Supervisory Section)

Energy is the capacity for performing work. A mass raised against gravitational force possesses 'potential energy' because it has a potential to move by virtue of its position. An object in motion possesses energy by virtue of its motion and that energy is called kinetic energy.

Potential and kinetic energy are both forms of stored energy relating to the potential to move or the motion of bodies and as such, form the basis of mechanical physics. There are, however, many other forms in which energy exists. Oil is inert when left to itself, yet it contains latent chemical energy which can be used in diesel or petrol engines. Energy is stored in the magnetic field of a permanent magnet and will do work when it moves iron brought into the vicinity of the magnetic field. Also, energy can be transferred from place to place in the form of radiation, which can be heat radiation or light, in fact, any form of electromagnetic radiation and it is in that form that the Earth receives energy from the sun.

Einstein's Theory of Relativity modified the theory on the conservation of mass, which states that matter can be re-arranged but not destroyed. The equivalence of mass and energy has been verified and is illustrated by the phenomena of radioactivity, the explosion of the atom bomb and the production of energy by nuclear fission. When the fission or fusion process takes place an amount of energy is released, relative to the change in mass. If the above technology had not been developed and put to work for the benefit of mankind, the population of the world would now be awakening up to the fact that at the present rate of use of energy, all the worst manifestations of an energy shortage would be on the horizon for about the year AD 2000.

To illustrate the point made above, a human being needs only the equivalent of 0·15 kW per man as food for effort (1 kW = 239 calories/second), but European man uses 5 kW/per man, both for food and other uses, to provide the material standard of life to which he is accustomed. In underdeveloped countries the figure is about 0·5 kW/per man, while in the USA the figure is 10 kW/per man. By the year 1980 the world will require energy at the rate of about 13×10^6 megawatts (ie the equivalent of 13,000 off 1,000 megawatt Power Stations), but if the whole world were to be brought up to the USA levels of consumption, the above figures would have to increase by a factor of six, ie the annual rate of energy consumption would be 78×10^6 megawatts, which if it was all fuelled from the world's present oil reserves, could be sustained for $5\frac{1}{2}$ years.

Looking at the energy scene in another way, if the growth demand continues at about the same average annual rate of 3·5 per cent, as in the last 50 years, between the years 1975 to AD 2000, 2·4 times more energy will be consumed in these 25 years than in all previous time to the year 1975.

All life on earth is dependent on the element carbon. In general, all animals, ie organised being s having life, feeling and the power of voluntary motion, fuel that life with carbohydrates, proteins and fats. The main source of carbohydrates is green plants, which through the photochemical assimilation of carbon dioxide from the air produce carbohydrates. All life is dependent on the operation of this process, which is in effect the conversion of the Atomic Energy of the sun (ie Nuclear Fusion) into forms suitable for the chemical transformations of living matter.

As stated above, all animal life uses the carbohydrate food chain to sustain the physical life, but the higher order of animals, namely man, to enable him to live in the present day highly organised life style, has had to start using hydrocarbons to produce the energy needed to operate the machinery and heat his home, thereby giving him the standard of living he wishes to enjoy. That activity was acceptable when the world population and the standard of living was much lower than it is today, but the rate of use of hydrocarbon fuels has now reached such a high rate that the world faces a fossil fuel shortage by the year 2000 (for oil) and 2150 (coal). Allied to the above is the problem of feeding a rapidly increasing world population which requires a massive energy input to provide fertilizers in sufficient quantity to provide the food that will be required, just to feed the world's population at the present level of consumption. If the underfed two-thirds of the world's population insists on having as high a standard of living as the Western world, as they surely will, that again creates a demand for more energy.

Opponents of the expanding use of nuclear power advocate the development and expansion of renewable energy sources such as hydro-electric schemes, geo-thermal power stations, tidal power schemes, wave power, windmills, solar power schemes, etc. The only problem with these proposals is that if every one of these energy production schemes were put into effect, their total contribution to the energy needs of Britain would equal about 6 per cent, or at the very outside, 10 per cent. That leaves 90 per cent plus to be provided by other means, which in the light of the fact that hydrocarbon fuels should be being preserved for the use of future generations as chemical feedstocks, leaves nuclear power as the obvious alternative to that of wastefully burning coal and oil.

The present technology of nuclear power uses the fissioning of uranium to produce heat which in turn produces steam which is used to drive a turbo-alternator to produce electricity. The above technology is being used in Britain in 'Magnox' nuclear power stations and the advanced gas-cooled reactor power stations and it is intended to use a similar fissioning process in the steam generating heavy water reactor, which is currently being designed at a commercial rating of 660 megawatts. All the above types of reactor produce plutonium, which is the most toxic substance known to man, but the benefit that accrues to manking is the fact that plutonium can be used in reactors known as Fast Breeder Reactors, which produce more fuel than they burn. As the demand for energy increases, so will the demand for uranium, but just as coal and oil reserves are finite, so is the amount of uranium that is available at $10 to $30 per pound, which are reckoned at present day valuations to be the minimum and maximum acceptable prices. It is the above situation that demands, that if Britain and the world are going to be self-sufficient in energy production, they will have to develop the Fast Breeder Reactor as a commercial proposition, if they are going to survive in an energy hungry world. Additionally, there is a requirement to develop the High Temperature Reactor (HTR) both for the generation of electrical power and as a source of process heat for the conversion of coal and oil feedstocks into other more useful chemicals. Allied to the above use of the HTR is the possibility of linking the process heat to the Eva and Adam heat conversion process which could then be used for district heating.

In America the opponents of the use of nuclear power put up all sorts of esoteric arguments to justify their opposition to an energy source that is in urgent need of development, if the American people are to maintain their high standard of living. To put the present use of nuclear power in America into its proper perspective, America at present produces more power through burning wood than it produces from all the nuclear power stations currently in operation.

For those people who are interested in statistics, the notes shown below taken in conjunction with the points made above, will enable them to calculate their own assessment of the world population/energy problem.

Population

The world's population is now growing at the rate of about 2 per cent per annum, which implies a doubling of the population every 35 years. In mid-1970 the Earth's human population passed 3,600 million and the 1974 distribution is as shown below:

World population		*Population of some major countries*		
Africa	393,764,000			
Asia	2,269,072,000			
		China	=	824,961,000
		India	=	586,266,000
		Japan	=	109,671,000
		Indonesia	=	127,586,000
Europe	660,418,000			
North America	336,499,000			
South America	208,315,000			
Oceania	20,719,000			
USSR	Totals included in Asia and Europe	Asia		63,255,000
		Europe		188,809,000
World total	3,888,787,000			

Population growth and projection

In 1750 World Population = 800 million

In 1950 World Population =1,950 million

In 1980 World Population will be over 3,900 million

At 2 per cent current growth rate, the world population will be approaching 8,000 million by the year 2010 AD.

World energy reserves and resources

The world energy quantities are very large, therefore the units are quoted in terms of Q:

where 1 Q $=10^{18}$ British thermal units

$=1{,}000{,}000{,}000{,}000$ million Btu

$=1{,}055\times10^{18}$ Joules or $1{,}055\times10^{18}$ watt-seconds

$=185\times10^{9}$ barrels of oil (661)

$=36{\cdot}6\times10^{3}$ million tonnes of coal equivalent (mtce)

The World Energy Conference 1974 developed the undernoted definitions on Resources and Reserves of Energy:

1 *Resources* of non-renewable raw material are the total quantities available in the Earth that may be successfully exploited by man within the foreseeable future.

2 *Reserves* are the corresponding fraction of resources that have been carefully measured and assessed as being capable of exploitation in a particular nation or region under present local economic conditions using existing technology.

3 *Recoverable Reserves,* are that fraction of the reserves-in-place that can be recovered under the above economic and technical limitations.

4 *Additional Resources,* embrace all other resources not yet discovered, but believed to exist from geological evidence.

5 *Renewable Resources,* such as hydraulic and geothermal, must be treated differently as they are rate controlled rather than stock resources. These resources are frequently quoted in terms of an integrated annual energy value derived from a maximum power level (in megawatts).

Estimates of available energy resources are a function of the state of technology at the time of estimation so that projections into the long term future can only be forecasts subject to wide margins of error. Nevertheless, by assuming today's technology and by predicting some future advances in technology from known scientific research, it is possible to indicate in the Table shown opposite, the variety and extent of resources which could be used over the next 50 years.

In the Table below 'Recoverable Reserves' show a total energy availability of 49Q. Between now and the year AD 2000 the world is expected to use around 11Q of the Recoverable Reserves. That would leave 38Q if no new 'Resources' were developed for future use. The problem that energy planners have to face is that by the year AD 2000 the world will be using the Recoverable Reserves at the rate of 0·82Q per year, which indicates the necessity to develop the 'Resources' and more especially the 'Potential Resources' which could enhance the 'Recoverable Reserves' by 200Q if the Fission (Fast Breeder Reactor) is developed.

World primary energy reserves & resources

(Units Q $=10^{18}$ Btu $=1055\times10^{18}$ Joules)

	Recoverable reserves	*Resources (estimated)*
Coal	35	325
Oil	6	25 (Oil and Natural gas)
Natural gas	6	
Fission (thermal reactors)	2	10
Total	49	
Hydro (-/annum)	0.1	
Potential resources		
Fission (Fast Breeder Reactor)	200	4,000
Fusion (Deuterium-tritium)	240*	1,000
Fusion (Deuterium-deuterium)	100×10^{6}	$10{,}000\times10^{6}$?
Solar (-/annum)	0.5	1,000 (total earth surface)
Geothermal (-/annum)	10	1,000

*Limited by lithium supplies of recoverable nature

Source: Hubber, M. King. Energy needs and the environment, *United Nations 'J' Series, World Energy Supplies* 1961-70

Paper submitted by the National Executive Committee of the Labour Party

ENERGY

The last three years have seen far-reaching changes in the world energy scene—changes which have had a major impact on consumers and supply industries in Britain. The days of cheap imports of Middle East oil have come to an end and the need for planned, long-term development of Britain's own energy sources, urged for many years by the Labour Party and trade unions, is now generally accepted. The hardships caused to consumers, especially low-income families and pensioners, by the sudden rise in costs of heating and lighting also make it clear that as a community we must plan to ensure that in future every home is equipped to use energy efficiently and that every household can afford to maintain a comfortable standard of warmth and light.

The need for long-term security, as well as the immediate problems created by unco-ordinated development of the pattern of energy supply and use, imply that we must urgently develop the means to plan the energy sector effectively and to ensure democratic accountability of energy planning to the taxpayer, consumers and workers in the supply industries. Our objectives are to ensure that:

1 everyone can afford adequate heat and light at home;

2 industry's needs for energy are fulfilled at a price which reflects full resource cost and has regard to the long-term availability of the various fuels;

3 these objectives are met on a long-term basis, taking account of risks; the depletion of our reserves of oil and gas is regulated; research and development in energy supply and use; is adequately funded; and investment in energy industries to meet those objectives is properly planned;

4 freedom of the consumer to choose between fuels provided at a minimum price which reflects economic cost should where possible be maintained and increased.

A NEW PLANNING FRAMEWORK

The Government must take responsibility for planning our energy policy. The statutory and administrative framework should be amended to permit this:

1 legislative provision must be made to give the Secretary of State power of specific directive over nationalised fuel supply industries in pursuit of this policy;

2 the statutes of the nationalised industries must be amended to bring them into conformity with the controlling assumptions of the policy;

3 the Secretary of State should have a duty to approve the long-run planning objectives of each supply industry, and not simply the investment programme;

4 the statutory financial obligations of the nationalised industries will have to be altered to allow for possible changes in the pattern of spending.

Accountability and planning

There is a need for the community to be involved in the process of planning our energy policy. This should include:

1 an energy forum with a secretariat and research unit in which representatives of industrial management, unions and consumers have the right to full information and consultation on a regular basis before the Government reaches decisions. This body should be established by statute and the public corporations, including the British National Oil Corporation, should have a statutory duty to co-operate with it;

2 a statutory obligation upon the Secretary of State to publish a regular policy statement for presentation to Parliament;

3 publication of a comprehensive review of energy prospects with all the alternative options for supply and conservation fully set out and costed;

4 corporate planning of each of the energy

supply industries on a fully tripartite basis, as carried out in the coal industry in 1974, reflecting the need for greater industrial democracy;

5 controls on competitive advertising by the nationalised energy industries and more information provided by Government itself about the costs of alternative fuels and heating systems.

OIL

By the end of 1976 we should be meeting about a fifth of our national oil requirements from the North Sea. The revenue from this resource must be used to assist the general development of all the regions of Britain; and in particular, North Sea oil and gas will mean a great future for the petrochemicals industry.

Participation

In our October 1974 Manifesto we were committed to 'take majority participation in all future oil licences and negotiate to achieve majority state participation in existing licences'. The Government is now negotiating with the oil companies to buy 51 per cent of the oil under existing licences. It will also purchase, as of right, 51 per cent of the oil under future licences, at a 'fair' price.

We believe, however, that wherever possible participation should involve a majority stake in the oil fields themselves. We recognise the difficulties in renegotiating the existing licences, granted under the Tories. But in future licensing the public sector's stake in development and production must be 51 per cent at minimum and up to 100 per cent where appropriate. We believe that it is not at all appropriate for this type of investment to be subject to ordinary public expenditure limits. It represents investment in new, highly profitable productive assets. If necessary, we would expect the Government, or preferably, our own oil company, the BNOC, to raise capital by borrowing abroad specifically for this purpose.

British National Oil Corporation

The British National Oil Corporation (BNOC) must be encouraged to build up its expertise and resources as rapidly as possible: and we welcome the decision to take over Burmah Oil Development. The BNOC has also purchased shares in the Ninian and Thistle fields in the North Sea formerly held by Burmah Oil. We believe that BNOC should become the sole licensee in promising sectors for which licences are granted under future rounds. In addition, its activities should be extended as rapidly as possible into downstream activities, particularly refining. The Government should give the Corporation a specific direction to do so.

British Petroleum

The role of British Petroleum must be re-examined in the light of the present public stake which now stands at almost 70 per cent of the total equity. The BP shares formerly owned by Burmah Oil, should be retained by the Government in order to strengthen its control and to ensure that British Petroleum acts in future in a way that is consistent with Government policy. It should be brought firmly within the public sector and treated as a publicly owned and publicly accountable company. At a later stage it may be appropriate to bring it within the aegis of the British National Oil Corporation.

COAL

We believe that coal will make an increasingly important contribution to Britain's energy requirements in the foreseeable future. The jointly agreed Plan for Coal to generate some 42 million tons of new capacity by the mid 1980s, should not be delayed by the temporary market recession. Investment in new capacity must continue. We welcome the decision to develop the Selby coalfield. The establishment of new coalfields will present enormous challenges of adjustment in the areas concerned for the miners who will work there. If these problems are handled successfully, however, a valuable contribution will have been made to the pattern of social development.

In the longer term we must concentrate on expanding the market for coal by encouraging its use in power stations, building new coal fired power stations, converting oil fired stations where possible, and, allowing for the coking coal shortfall, eliminating imports of coal. We would also encourage the direct use of coal in industry, and the development of new markets such as substitute natural gas production and district heating. We believe also that there is great scope for the use of a wide range of grades of coal to provide alternative heating systems within towns and cities with the development of the fluidised bed combustion system.

NUCLEAR POWER

By 1980, about a fifth of the electricity generated in Britain will be by nuclear power. We believe that the decision to opt for the British steam generating heavy water reactor for the next generation of nuclear power stations was the correct one and that we have achieved for the time being a reasonable balance in respect of nuclear power.

Although so far the safety record of our nuclear stations has been outstanding, we are conscious of the possible risks involved to the people and the environment, in the continued extension of nuclear electricity generating capacity. Our energy plans must take into account the full social and economic costs of safety needs in the future. We believe in full and open public debate on this issue, especially in regard to the processing and storage of spent nuclear fuel which can affect the lives of future generations as well as our own. We urge that the highly successful state-owned company, British Nuclear Fuels Limited, under the supervision of the statutory Nuclear Inspectorate, should give close attention to all these matters in the public interest.

FAST BREEDERS AND FUSION

In order that potential risk from reprocessing is kept to a minimum and that Britain is not caught in a possible uranium supply shortage, we support the development of the Fast Breeder Reactor, improving fifty-fold the ultilisation of uranium.

We therefore welcome the continuing work on the prototype 250 MW fast reactor at Dounreay and take pride in the fact that Britain is in the forefront of technology in this field. But there is a real question of how far Britain should go it alone on this expensive project. There is a good case for joint European effort into the research and development.

We believe that work should also continue on the production of energy from nuclear fusion and that the Joint European Torus (JET) is probably the best way of achieving this. Britain has more than fifteen years experience in this field through the work of the Atomic Energy Authority at Culham. Consequently we are of the opinion that joint international research and development on JET should be carried out at the Culham Site.

ELECTRICITY

We appreciate that increases in the price of fuel supplies to the power stations results in an increase in electricity prices to the consumer. Therefore we support the efforts to increase thermal efficiency at the power stations. There is a need for more coal-fired power stations to be built; our relatively scarce oil and gas resources should be used in other ways—as chemical feedstock for example.

We note the report of the Plowden Committee of Inquiry into the structure of the electricity supply industry but we are sceptical of some of its proposals and will continue to study them closely. We accept, however, that there is a need for change in the structure of the industry.

GAS

Natural gas is a valuable source of energy but the reserves will have a relatively short life. The available gas therefore, should go to those markets which can use it most efficiently.

Depletion policy is of particular importance. For the immediate future gas will probably be under priced in comparison with other fuels, but higher domestic prices might damage gas in the kind of market that can best use its properties as a highgrade, clean and convenient fuel. Gas from the Frigg and Brent fields has already been purchased. Damaging the domestic market may thus simply ensure that those supplies have to be diverted into the non-premium industrial market.

The use of natural gas in power stations, and to provide crude heat to industry, can only be justified to provide the industry with flexibility in the face of sudden surges in supply and demand, and to ensure a reasonable loading on the transmission system.

We support research into methods of producing substitute natural gas from coal to keep this important option open, if necessary to plug a potential energy gap in the 1990s. We also support the experiment into methods of converting natural gas into an edible protein.

ALTERNATIVE ENERGY RESOURCES

The increase in the cost of fossil fuels has awakened an interest in the development of new energy sources. We welcome the growth of expenditure into alternative schemes, although this expenditure is still too low. Some of these sources, such as geothermal energy, are not even at the experimental stage. But we support the extension of research and development into this field, and also into solar energy. Wind energy too could be used as a marginal source of energy in the electricity generation sector. We support the study into the use of wave and tidal power, particularly the Severn barrage scheme, and the extension of district heating systems which use waste heat, instead of fuels, from power stations and incinerator refuse. It is through the use of alternative energy sources that we can achieve maximum fuel conservation and significant savings in our imports.

CONSERVATION

Although the Government has undertaken a number of short term measures to encourage energy conservation, more urgent and stringent action is needed if we are to make any worthwhile savings. For example, there are advantages to be gained from investment in new housebuilding and the alteration of existing stock to improve standards of insulation in such a way as to drastically cut the energy needed to maintain warmth. We should set ourselves a target, say, over 25 years, to bring households up to the best insulation standards. We also suggest that in order to give the consumer a choice between forms of domestic heating the practice of building new houses without flues should be re-examined. Building regulations and subsidies must be tailored to this programme. Before regulations are finalised there must be extensive consultations with tenants, owner-occupiers and industry. The Advisory Council on Energy Conservation, set up in October 1974, must assist the Government with the preparation of reports on this question.

INTERNATIONAL ENERGY AGENCY

Following the 1973 oil crisis the major industrial nations recognised the need for international collaboration in the more efficient use of available energy resources. In the developed world, Britain in particular will be in a strong position of surplus in coal and oil and we will have an important role to play in the international energy field. Britain has a duty to act responsibly towards the developing world's energy needs in the future.

ENERGY PRICES

A cheap energy policy is not possible because of the budgetary cost of holding down coal, electricity and gas prices. But if energy prices are to remain high we believe it is the Government's duty to make financial provision to ensure that low income families can afford a reasonable standard of warmth.

At present the price of our oil is fixed by the world market while prices of coal, electricity and gas are mainly determined by the financial targets set for each industry. This has caused large, sudden and distorted increases in prices of different fuels. It is wrong for prices to be determined by short term financial objectives, since the profit and loss of each industry varies from year to year on account of fluctuations in capacity utilisation, interest rates, etc. Instead, the price structure should be related to the long-term objectives of energy policy. The financial targets of the energy industries should then be made consistent with this pattern of prices.

Statement by John Biffen, the Conservative Party spokesman for energy

This statement is given in Volume 1 of this Energy Paper, *Report of proceedings; written contributions.*

Liberal Party submission: electric heating and council tenants

As long as the bulk of electricity is generated by combustion of hydrocarbon fuels, heating by electricity is certain to be more expensive and wasteful than direct heating by combustion of the same fuels, either in individual domestic heating systems or in direct heating schemes. In practice the upper limit of thermal efficiency for electricity generation is about 35 per cent, and further losses occur in transmission and distribution. Thus two-thirds of the energy content of the primary fuel is lost in the cooling water and flue gases at power stations, and of the remainder a proportion is then dissipated in resistance losses in the transmission and distribution network. With modern domestic appliances, on the other hand, more than 70 per cent of the energy content of the primary fuel remains as useful energy in the home.

The gain in efficiency from direct use of primary fuels for domestic heating is partly offset by the higher capital cost of installation if gas , oil or solid fuel is used compared with the equivalent for electricity, but at existing price levels per unit of useful heat, electricity is still far the most expensive form of heating. Nevertheless in 1974 local authorities still chose electricity for as many as 26·1 per cent of dwellings in tenders approved during that year,[1] and it is estimated that 700,000 tenants are now living in all-electric heated dwellings.[2]

The Department of the Environment has no policy on the type of heating adopted for new dwelling by local authorities. Provided the schemes submitted to the Department for loan sanction conform with Parker Morris standards and do not exceed the housing cost yardsticks, the central government has no power to influence the choice of fuel, and there has been no attempt to do so indirectly by calling attention to the energy conservation implications of the decision. If anything, the cost yardstick requirements tend to favour electricity, since cost in use calculations do not allow for the inevitable price increases that will occur in the future.

Assuming that the cost yardstick procedures are retained, local authorities should be required to submit alternative estimates where the first choice of fuel is electricity, giving the extra capital cost of using another fuel, and the calculated annual saving to the tenants.

The National Economic Development Office (NEDO) has estimated that for a new two-storey semi-detached house with an area of 920 sq. ft, the annual net energy demand is 67·7 GJ.[3] Mr Neville S. Billington, former Director of the Heating and Ventilating Research Association, estimates that for a dwelling of 100 m^2 floor area (1,076·4 sq. ft), the net annual energy use, allowing for a mean house temperature of 18°C, would be 64 useful GJ.[4] *Which* suggests that 79·2 GJ for 'a typical three bedroomed house' area unspecified[5]. These figures, although of the same order of magnitude, illustrate the difficulty of making comparisons of heating costs on an average basis. The net consumption will vary considerably according to the floor area of the dwelling, the type of construction (terrace, semi, intermediate floor flat, top floor flat), the mean house temperature and the standard of insulation. The end terrace or semi-detached dwelling, for instance, has a heat load some 45-50 per cent higher than the centre terrace[6]. Taking the three-bedroomed semi-detached house as an example, the Director of Mechanical and Electrical Engineering of the GLC estimated the cost of electric heating in block heaters as £207 per annum, and of gas heating, £149 per annum[7].

In the case of tenants receiving supplementary benefit, any additional rent that might be payable as a result of the higher first cost of gas, oil or solid fuel heating would be recoverable, whereas the extra cost of heating by electricity has to be borne by the consumer. The Supplementary Benefits Commission (SBC) only take the average cost of energy into consideration in determining the scale rate allowances, and the GLC figures show that a difference of as much as £1·12 a week can exist between the actual cost for two households with precisely the same accommodation and standard of heating. In fact, the variation would be even larger if differences in architectural form and insulation were taken into consideration.

The standard amounts adopted by the SBC for heating in specified benefit calculations are some indication of the average amount a national average pensioner might be expected to spend. In June 1973 these amounts were £1·20 where heating only was involved and £1·35 where all full requirements were included. The last increase was in November 1975, when the amounts were raised to £2·20 and £2·50 respectively. During the same period the average revenue per unit from sales of electricity to domestic consumers in England and Wales rose from 0·97p/unit (quarter ended 1973 June 30) to £2·28p/unit (quarter ended September 30, 1975). Therefore if a consumer was spending £1·35 on electricity in the second quarter of 1973, the same number of units would have cost him £3·17 in the third quarter of 1975, compared with the £2·50 allowed. Putting it another way, if the pensioner was to have the same amount left to spend on items other than energy, he would have had to reduce his consumption of electricity by 21 per cent between the two dates.

The conversion of dwellings from electricity to gas, and the substitution of gas for electric water heating, should be considered for estates where gas mains have been laid. But in the meanwhile, the burden of heating costs on tenants with electric systems can be reduced by improving the insulation. For the winter of 1974–75, the Institution of Heating and Ventilating Engineers estimated that for a bungalow of 1,000 sq. ft, ceiling area, the saving from installing 4 inches of roof insulation would be £106·30 while the cost of material and professional insulation would be £122. The savings were calculated on an assumed electricity cost of 0·84 p/kWh, while this winter the average actual cost has been in the region of 0·95 p/kWh.

Under the Circular 160/74, 'Improvement of Older Housing', local authorities are encouraged to apply for grants—up to 50 per cent of approved expenditure—in respect of dwelling occupied by elderly or disabled persons, where thermal insulation of roof spaces is to be brought up to the standard limit down in Part F of the Building Regulations.[8] Up to the time of writing, no such applications had been made, either for owner-occupied dwellings or for those owned by the local authorities themselves. The red tape involved is such that it is not surprising that no elderly or disabled owner occupiers have applied, but local authorities, which have the staff to deal with the necessary paperwork, seem to have overlooked the money which is available to householders under the provisions of this circular. As the Minister at the Department of the Environment has said in a letter dated March 29, 'We suspect, however, that authorities may not all be aware of the position, and we intend to give further publicity to the arrangements in a circular we shall shortly be issuing, concerning this and a number of other questions which have arisen over subsidies for improvement works.'

Since, furthermore, the labour needed to put in loft insulation is unskilled this would be an ideal way of using unemployed school leavers under the job creation programme. A determined effort on this front would also yield a significant energy saving, as a contribution towards the national conservation programme. Finally, it would reduce the local authorities' own maintenance expenditure, by reducing condensation in the dwellings concerned.

For these reasons, we recommend that local authorities should be obliged to bring up to the standard laid down in Part F of the Building Regulations the insulation of roof spaces in all dwellings owned by them and without any form of heating other than electric. If this were mandatory, no grant incentive would be necessary, but the local authorities should be permitted to recover their expenditure by increasing the rents over the remaining life of the property. There might have to be an exception in the case of dwellings with less than, say, five years life. Tenants in receipt of supplementary benefit would recover the additional rent from the SBC as well as gaining from lower fuel bills. For tenants in employment, the lower fuel bills would more than compensate for any increase in rent.

During the year 1975–6 the Electricity Council, while not promoting electric heating as such, ran advertising campaigns designed to influence consumer opinion in favour of storage radiators. In August and September a press advertising campaign was launched at a cost of £75,000, to emphasise that off-peak tariffs were not being phased out. To augment this message and to link it with the desirability of thermal insulation, a TV campaign occurred in September at a cost of £100,000. In October a further £50,000 was spent on TV advertising of storage radiators coupled with an offer of price reductions on thermal insulation material.

Advertising by the rival nationalised fuel

industries does not help the consumer to make rational choices between them. On the contrary, it is extremely difficult for even the most determined consumer to evaluate the differences in capital and operating costs when the industries use different units, and their tariffs are constantly being altered. There is a need for a central unbiased information agency, to advise individuals, local authorities and private house builders on these matters, which could be financed by means of a levy on the gas, electricity, and coal industries. If their own advertising budgets were reduced by the amount of the levy, such a scheme could be introduced without any cost to the consumer.

BIBLIOGRAPHY

[1]Written Answer: House of Lords 22/3/76

[2]*Energy Tariffs and the Poor* Dept. of Energy 1976

[3]*Energy Conservation in the UK – Achievements, Aims and Options* NEDO 1974 p 29-30

[4]*Thermal Insulation Standards*, N.S. Billington, Pergamon Press 1975

[5]*Which* report September 1975

[6]BRE Paper on Energy Conservation: June 1975

[7]Information from the GLC

[8]DOE Circular 160/74.

Printed in England by Harrison & Sons (Hayes) Ltd. and published by Her Majesty's Stationery Office.

Dd 497261 K20 7/76